THE SPARK O

THE
SPARK
OF LIFE

Darwin and the Primeval Soup

CHRISTOPHER WILLS
JEFFREY BADA

PERSEUS PUBLISHING
Cambridge, Massachusetts

Every effort has been made to obtain permission to reproduce material in this book. Many of the pieces of art were collected with the generous assistance of the National Library of Medicine.

The authors are also grateful for the help of the National Aeronautics and Space Administration in publication of this book.

A CIP catalog record for this book is available from the Library of Congress.
ISBN 0-7382-0493-5

Perseus Publishing is a member of the Perseus Books Group.

Find us on the World Wide Web at http://www.perseuspublishing.com

Perseus Publishing books are available at special discounts for bulk purchases in the U.S. by corporations, institutions, and other organizations. For more information, please contact the Special Markets Department at HarperCollins Publishers, 10 East 53rd Street, New York, NY 10022, or call 1-212-207-7528.

Text design by Cynthia Young
Set in 11-point Minion by the Perseus Books Group

First paperback printing, February 2001

1 2 3 4 5 6 7 8 9 10—03 02 01

Dedicated to Stanley Lloyd Miller,
colleague and friend,
whose half century of devotion to the question of the origin of life
has transformed the field from speculation to hard science

CONTENTS

ACKNOWLEDGMENTS

WE ARE MOST GRATEFUL FOR THE ASSISTANCE OF MANY DIFFERENT kinds, scientific and logistical, that we have received from Gustaf Arrhenius, Bill Baity, Oliver Botta, David Brin, Mike Clark, Steve Freeland, Michael Gaffey, Danny Glavin, Susan Green, Jenny Hamilton, William Hartmann, Gerhard Kminek, Keith Kvenvolden, Lois Lane, Antonio Lazcano, Marilyn Lindstrom, Norman Pace, Karsten Pedersen, Glenn Schneider, Bill Schopf, and Susan Terebey. Special thanks to the students and postdoctoral fellows who participated in our Origin of Life discussion seminar in 1998. David Deamer, Russell Doolittle, Walter Fitch, Antonio Lazcano, and Stanley Miller read and commented on parts of the manuscript, catching many problems in the process, but of course we shoulder the responsibility for any errors of fact or interpretation that remain. We are also indebted to our editor, Amanda Cook, for pulling us up short whenever we tended to stray into thickets of complexity. The University of California, San Diego, Chancellor's Associates and National Aeronautics and Space Administration Specialized Center of Research and Training in Exobiology provided support.

INTRODUCTION

THERE IS A PENINSULA IN SOUTHERN SWEDEN THAT EXTENDS into the Baltic and is surrounded by an archipelago of beautiful wooded islets. On it looms an immense dinosaur. This dinosaur is not like the ones in *Jurassic Park*. Rather it is a soon-to-be extinct nuclear power plant that provides more than 10 percent of Sweden's power. The plant, along with Sweden's eleven other nuclear power plants, is scheduled to be shut down permanently by the year 2010, a deadline demanded by a referendum that the Swedes voted into law in 1980. But where to put the toxic radioactive waste that these plants have generated?

This problem has inspired a huge underground experimental instal-lation at nearby Äspö. There, a gently sloping tunnel three kilometers long has been bored into the hard Swedish granite, gradually dropping down to a depth of five hundred meters. At the bottom of this tunnel, in a series of caverns and boreholes, measurements are being made to determine how quickly water can trickle down from the surface to this depth. These and other measurements will help to establish the rate at which the huge copper cylinders that are designed to encase the ra-dioactive waste will corrode away.

The cylinders must survive for at least a hundred thousand years. Free oxygen is likely to accelerate the corrosion process. Yet the excava-tors have found, to their surprise, that when they inject oxygen into the rock, the gas rapidly disappears or is converted to carbon dioxide. It seems that bacteria are at work here, far below the surface.

Periodically, the excavators break through into one of a number of mazes of underground water channels. And when they do, Karsten Ped-ersen of the University of Goteborg is on hand, ready to sample the microbes living in the water. He has found a great variety of different bacterial species, most of them anaerobic or able to survive only in the presence of minute amounts of oxygen. Here, far below the surface, they carry out many of the same chemical reactions that are performed by their relatives far above. A community of microbes thrives in this unlikely place, with little obvious connection to the surface.

Pedersen wonders whether this is the place, deep within the crust of the Earth, where life could have begun. The rock faces certainly look like very unpromising sources for life. The surfaces from which the samples are taken are black and enigmatic, scored with the traces of the drilling and blasting necessary to smash through the iron-hard rock. They are filmed with condensed moisture and studded with the gauges and stainless steel tubes that measure water flow and gas diffusion. Now that the drillers and excavators have left, this underground world has reverted to a quiet darkness. There is no doubt that before humans and their machines arrived on the scene, the tiny inhabitants of this subterranean world lived lives of profound uneventfulness.

Is Pedersen right? Could life have had a subterranean origin? The samples taken from the rocks have given no clear answer to the question. The microbes that he has isolated and grown are strikingly varied, with representatives from three large groups of bacteria. They exhibit a wide range of chemical abilities. And, even though they now make their home in these deep rocks, they probably did not arise there. They are much more likely to be relatively recent arrivals from above, filtering down with the slow creep of water through hairline cracks in the rock.

Of course, early in our planet's history, the underworld was a more active place. It was heated from below, and hot water geysers fractured the rocks. The resulting cracks were filled with plentiful quantities of gases like hydrogen and methane, working their way up from vast reservoirs of gas far below. But the subterranean regions would still have been relatively peaceful compared to the tumult on the Earth's surface above.

Like Pedersen, many scientists are entertaining the notion that life could have first appeared in such quiet subterranean depths. But we disagree. We think, and we hope to persuade you in this book, that such a dark and dull place is an unlikely place for life's beginnings. We will present strong arguments that life first appeared on the Earth's surface, and that its emergence was driven by active processes. Tides, wave action, and periodically erupting geysers, along with flashes of lightning and blazes of sunlight and ultraviolet radiation from above, were continually altering the environment and providing strong, immediate, and specific selective forces that aided the emergence of life. In short, we think that the fierce pressures of Darwin's natural selection played a central role in the origin of life almost from the beginning and that life would never have appeared without them.

Organisms are still at war with their environment today. The very first living entities, and indeed even those almost-living collections of molecules that must have proceeded them, were certainly locked in a struggle with their environment as well. It is the source and nature of this battle that concerns us here.

Where Did Life on Earth Come From?

This book is about some of the greatest and most difficult questions that still confront science. How did life originate on Earth? Did it appear more than once? Is there life elsewhere in the universe? And, perhaps the most dramatic question of all, do we have it in our power to duplicate the origin of life in the laboratory?

Some would say that this is an impossible quest, far more challenging than the other major scientific problems that currently face us. For example, one of the great tasks confronting science in the new millennium is to determine how the brain works. Daunting as this task is, at least we can do experiments on brains. But how can we reconstruct, in a convincing way, the world as it was when life emerged four billion years ago? That world was utterly different from ours, and we are still ignorant of much crucial information about it.

Only if we can perform such a reconstruction can we design experiments that will allow us to create life in the laboratory: the ultimate— and indeed the only—satisfactory way to answer the question of how life arose. Surprisingly, as we will see, we are making substantial progress toward this remarkable goal.

Our puzzlement about the origin of life has itself had a long history. Most religions have of course attempted to answer it, and the result has been an astonishing variety of contradictory tales of creation. All tend to reflect the characteristics of the societies in which they originated. One of the best known is the version described in the book of Genesis in the Bible, which is actually an amalgam of several different creation myths. Others are not so well known, at least in Western cultures.

The Babylonian creation myth, a reflection of a warlike society, relates that the gods began to quarrel among themselves from the very beginning. An arrow from the god Marduk split the body of his grand-mother Tiamat into two halves, creating the heavens and the earth. The Hindu creation myth takes a gentler and more ineffable turn, reflecting Hindu society. The Upanishads present two contradictory views, one in which there was "being" at the very beginning and one in which there was "nonbeing" from which "being" arose.

Many of these myths of origin suppose a transition from chaos to order, and indeed our current scientific view of the origin of life postulates the same thing. With the advent of the scientific method, of course, we can now pose the question in a more precise and testable way—though the transition from a religion-dominated to science-dominated worldview has sometimes come at a terrible cost.

The philosopher and mathematician Giordano Bruno (1548–1600) thought that the universe was infinite and filled with worlds inhabited by sentient beings (and implying an infinity of different religions). His ideas, which suggested that religion was relative rather than absolute, were of course anathema to the Catholic Church. He was at odds with the Church hierarchy through most of his life and was repeatedly accused of heresy. He was finally trapped by the Inquisition and died a terrible death by fire, but his writings and ideas helped to power the Enlightenment.

Science has now demonstrated, or is about to demonstrate, the validity of many of Bruno's ideas. Although the universe is not infinite, it is certainly very large. Recent astronomical observations show that it is filled with solar systems. Some planets in these solar systems must have given rise to living organisms, and such planets now appear to be so numerous that a minority of them must harbor intelligent life. The mechanisms that brought about these repeated origins of life were not the actions of gods or the result of mysterious transformations from chaos to order that affected entire worlds simultaneously. Instead, they were chemical and biological.

Nonetheless, the detailed mechanisms by which life originated continue to stump the best scientific minds. The origin of life is such a puzzle that many people have suggested that we should simply abandon our hunt for extraterrestrial life until we know more about life on our own world and therefore the kinds of life that might be found on other worlds.

Scientific ideas about the origin of life have come and gone and have sometimes come again. Nonetheless, we have managed to inch forward, and the stage is now set for some stunning advances. Here we will trace

THE SPARK OF LIFE

the story of the search for answers to the question of the origin of life as scientists are currently pursuing it. We will, in the process, wander from one end of the world to the other. We will venture into the depths of our planet and into the Stygian blackness of the deep ocean. At the end of the book, we will go even further afield, traveling through near and far reaches of the universe as we search for clues to life elsewhere. The intellectual story we will trace is a fascinating one, inspired by a growing knowledge of chemistry, biology, and planetary science, but still hampered by a frustrating core of ignorance. And it is this core that we hope to penetrate, far enough to see the outline of the true story.

Life's Beginnings

We must begin, as so many have done before us, by confronting a difficult question of definition. How, in essence, does life differ from the inanimate world? And in particular, what properties did life have when it first separated itself from the nonliving world?

Those early transition creatures from the dawn of life are often referred to as *protobionts* rather than living cells. They were certainly not living cells as we know them, and indeed they probably had few of the characteristics of living cells today.

The mechanism that is usually proffered for the emergence of the first protobionts is a process called chemical evolution. This idea was originally explored by a pioneer in the origin-of-life field, Aleksandr Ivanovich Oparin. Oparin suggested that collections of molecules were continually coming together in a prebiotic soup, and that the ones that persisted the longest would come to predominate. Somehow, this chemical evolution led to the first self-replicating entities, or protobionts, and once this had happened, biological evolution took over.

This is the scenario presented in textbook accounts of the origin of life, and this mind-set had dominated thinking in the field. But there are too many gaping holes in this story for us. Protobionts arose somehow from the collection of organic molecules in the Earth's early primordial oceans. But we know that even if one were to recreate in a laboratory flask a rich and complex approximation of an early Earth soup, sterilize it, and then let the flask sit on a shelf, it would sit there indefinitely without any sign of life—or indeed any sign of Oparin's chemical evolution. There has to have been a powerful mechanism behind the appearance of the first protobionts.

We propose in this book that what is missing in the chemical evolution scenario must have been provided by the master of the evolutionary process, Darwin himself. The current scenarios for chemical evolution force Darwin to wait impatiently in the wings, able to do nothing until the first protobionts appeared. But Darwinian evolution has strict rules, and at least some of those rules must have been operating even before the beginning of life. It is time to explore the evolutionary pressures that led to the first protobionts. We would like to replace the vague and rather mystical concept of chemical evolution with something much more concrete—and more testable in laboratory experiments.

The necessity for Darwinian mechanisms becomes vividly apparent when we consider how different the first protobionts were from the nonliving collections of molecules around them. Even at this early stage in the evolution of life, the differences were numerous.

First, the protobionts were able to make approximate replicas of themselves. Such an ongoing chain of reproduction ensured both the survival of their lineage and the continuation of their ability to reproduce. The key word here is approximate, for the replicas were probably not exact copies of their "parents." Because the replication was inaccurate, this ensured that each new protobiont was slightly different from those that had gone before. Variation from one protobiont to another meant that some of them were better adapted to their environment than others. Because some protobiont lineages survived while others did not, true Darwinian evolution could begin.

The living organisms of today make extremely accurate copies of themselves. Only a few mutational mistakes are allowable, since offspring filled with such heritable mistakes are unlikely to survive. The very first protobionts were not under such tight constraints, because, of course, they were not competing with organisms that were better replicators than themselves. All that was required was survival of the ability to replicate, no matter how slapdash that ability might have been. Accuracy of replication was a secondary capability that probably developed only gradually. Nonetheless, they could replicate.

Second, these protobionts were able to survive under the savage environmental conditions that must have characterized the primitive Earth, such as blasts of ultraviolet light from the Sun, extremes of temperature and pH, a total absence of free oxygen, and widely varying concentrations of salts and other minerals. The protobionts must have been tough enough to withstand such unpleasantness, even though they were much simpler in every respect than living cells today.

Third, protobionts were somehow able to draw energy from that un-promising environment, either through the use of energy-rich gases such as hydrogen or through the light from the Sun. This allowed them to make the energy-rich, or activated, compounds that were essential to the replication process. Protobionts must have developed the capacity to make these molecules very early, for they could not have relied for very long on the few such molecules that happened to be present in the primitive soup. The soup quickly became exhausted as the first hungry protobionts drained it of anything useful.

Today, the primary molecule that captures and redirects environ-mental energy in living cells is adenosine triphosphate (ATP). Even to-day, many bacteria that live in extreme environments can manufacture only very small amounts of ATP, which means that they can only grow very slowly. ATP was probably not the first energy-rich compound, but it is impossible to imagine life without a source of energy very much like it.

Fourth, the Grim Reaper must have put in an early appearance. Death is inseparably associated with life. Without death, there could have been no sorting-out of the more successful replicating protobionts from those that were less successful. Right from the beginning, the most effective replicators and energy-utilizers were able to survive. At the same time, the less efficient ones were somehow destroyed or simply decomposed and thus removed from the competition.

Death, like reproduction, is an essential part of Darwinian evolution, provided that it occurs nonrandomly. Indeed, even before the appear-ance of the first protobionts, a process or processes that mimicked death must have begun to play a role. We will suggest in the course of this book that this analogue of death must have been operating from the very beginning in the primeval soup, and that this process helped in the formation of the protobionts themselves.

Notice that this minimalist definition of the first life says nothing about the structure or appearance of these early protobionts. It also says nothing about the genetic material that they possessed, other than to specify that the protobionts did have enough biochemical machinery to replicate this genetic material. Furthermore, their genetic material must have been able to code for at least some information about the properties of the protobionts. Without such genetic machinery, true Darwinian evolution could not have taken place.

Thinkers about the origin of life have wrestled, with varying degrees of success, with the question of how the first organisms could have acquired these properties. These thinkers can be roughly divided into those who believe that genes came first and those who believe that life must have started with metabolism.

Simply put, metabolism is the collection of chemical pathways in the cell that start with simple compounds and make them more complicated, or that start with complicated molecules and break them down into simpler ones. Our cells carry out their various functions through metabolic pathways, like the engine of a car as it burns gasoline. But to make more engines, you need a blueprint and a factory—the equivalent of genetic information. Otherwise, like those ancient cars that cough along the streets of Havana, all of us would senesce and die in spite of all our metabolic efforts, without leaving any offspring.

The physicist Freeman Dyson, in his 1985 book *Origins of Life* (note the plural!), tried to come up with a convincing metabolism-first scenario. He suggested that the first protobionts were bags of molecules, floating in the primitive soup, that could grow by acquiring more molecules from their environment. In some of these bags, a few of the molecules became organized into metabolic pathways. One pathway might make a molecule that could take part in a second pathway, and that second pathway might make another kind of molecule that could take part in the first pathway.

Through such complex, cross-feeding reactions, the bags grew in size. Once they had grown large enough, they then split into two smaller bags (bag-ettes?), each of which had approximately the properties of its parent bag. The daughter bags then grew until they could split in turn. Dyson pointed out that the collections of molecules contained in the bags were not true genes. He suggested that genetic material as we know it appeared later, after these self-replicating bags had appeared.

It is possible, in the laboratory, to introduce a certain amount of self-organization into a collection of molecules. But no laboratory so far has been able to make sets of molecules that can organize into pathways and increase in number without the aid of genes. We think that Dyson's scenario is unlikely, though as we will see there are some interesting ideas that lurk among his collections of quivering bags.

Dyson's idea illustrates vividly the severe fragmentation of viewpoints among scientists who deal with the origin of life. Dyson and other scientists, such as Günter Wächtershäuser of Munich, David Deamer of the University of California at Santa Cruz, and Doron Lancet of the Weizmann Institute in Israel, are firm believers that metabolism must have come first. Another and much larger group of scientists, including Stanley Miller of the University of California at San Diego, Thomas Cech of the University of Colorado, and Leslie Orgel of the Salk Institute for Biological Studies, believe just as firmly that gene replication must have come first.

This second group is hardly monolithic. There are endless arguments among them—for example, what was the nature of the first genetic material? Was it ribonucleic acid (RNA)? Did some other simpler genetic material precede RNA? Is it possible to construct simple molecules capable of carrying genetic information in the laboratory?

Other questions abound. What could that first genetic material do? How clever was it? Could it have made copies of itself entirely on its own, the molecular equivalent of pulling oneself up by one's bootstraps? Is it really possible to construct a molecule in the laboratory, made out of RNA or something simpler, that is capable of such an astonishing feat? Even if such a self-replicating molecule could be manufactured, would it really be "living"? Where would it get the building blocks to make copies of itself? If there has to be some external source of building blocks, then does this mean that Dyson, Deamer, and others are right and that metabolism really did come first?

Arguments about the origin of life tend to have this strange recursive quality, circling maddeningly around the problem without ever quite coming to grips with it. We are going to try to break this circle.

Life at the Present Time

Today, life is so tenacious and omnipresent on the Earth that it is difficult to imagine the planet without it. If you wander over the fresh lava fields that have been extruded from a recent eruption on the big island of Hawaii, you will see spiders and insects making their homes in crevices of the barely cooled rock. It is not long before ferns begin to uncoil themselves hesitantly from those same crevices. Visit the ice fields of Antarctica during its brief summer, and you will find the larvae of midges in the pools of water that form on the surface of the ice. These insects, dining on blooms of algae, must race through their life cycle before the cold of winter sets in again.

In the artificial world that we have constructed for ourselves, we often try to exclude other living organisms, but always fail to do so. We think of ourselves as the dominant species on the planet, but any momentary inattention on our part always provides an opportunity for other species to reassert themselves. Bacteria and even insects thrive in the odious sumps filled with used crankcase oil that lie beneath decrepit service stations. In La Jolla, where we work, and along other parts of the California coast, harbor seals are multiplying again after decades of persecution. They are beginning to reclaim the beaches and to displace the beaches' human "owners."

Life as we know it is assertive, demanding, and unstoppable. It is driven by an enormous evolutionary imperative, one that blindly insists that the fittest organisms survive and the less fit fall by the wayside. Natural selection, though not formally a directional process, has led to greater and greater complexity in the living world over time.

Nonetheless, life had a beginning, and it is now possible to trace many events in the history of life almost all the way back to that beginning. If you examine the fossil record of half a billion years ago, you will find that there were no insects, no reptiles, no mammals, no flowering plants—indeed, few if any organisms that could live on dry land at all. Go back 2 billion years, a little less than halfway to the origin of the Earth itself, and you will find that the only organisms alive then were microscopic single-celled creatures that sometimes clustered together into slimy layers but showed little organizational ability otherwise. Go back 3.5 billion years, and . . . But that is what this book is about.

THE RISE AND FALL OF SPONTANEOUS GENERATION

[Francesco Redi's experiment] records the first, and therefore the most important, statement supported by experimental evidence of that great generalization named by Huxley the theory of biogenesis, a theory, which by its application, has probably been of more benefit to mankind than any other result of scientific investigation.

Mab Bigelow, translator's forward to Francesco Redi's Experiments on the Generation of Insects (5th ed., 1688, tr. 1909)

LET US BEGIN BY SENDING YOU, GENTLE READER, INTO THE past. Once there, you must forget everything you know. You must depend for your understanding of the world only on what was known at that time.

Imagine that you are a young person living in 1810, at the start of the second decade of the nineteenth century, and that you are just beginning your advanced studies. You have elected to learn about living organisms, their origins, and the relationships that they might have to each other. But you rapidly become confused. The more you learn, the more you find yourself tangled in a web of controversy and debate.

One of the great unsettled questions, you soon discover, concerns the phenomenon of spontaneous generation. Are living organisms being generated at the present time, spontaneously, from nonliving matter? It happens that you have begun your studies in the midst of a fierce argument about this question, an argument that has been going on for more than a century and that will continue for the better part of a century to come.

Although you do not realize it, the resolution of this problem will have the most profound consequences, ranging from the discovery of cures for infectious diseases to a deep understanding of evolution and genetics. But even when the matter is finally resolved, it will still leave unsettled the greatest puzzle of all, where and how life itself originated.

You will not live long enough to see the ultimate resolution of the spontaneous-generation question. Even so, you must try to make sense of the living world by using what your predecessors have discovered.

Some of the confusion that you encounter is a holdover from a time before the Dark Ages. The foundations of scientific observation and inquiry were laid down in ancient Greece and the Hellenized Mediterranean by Aristotle, Archimedes, Ptolemy, and other profound thinkers of the ancient world. Their early studies and speculations were inevitably filled with errors. Nonetheless, the scholars who rediscovered their writings during the medieval period were so dazzled by the brilliance of these early works that they raised them to the status of a kind of Scripture.

Aristotle stated that all kinds of living organisms could appear spontaneously, either in places that were becoming moist or those in the process of drying out. Authorities who wrote in medieval times, and even some who lived as late as the seventeenth century, agreed with Aristotle and claimed that mice, frogs, and eels could emerge from garbage, mud, and river water.

A Flemish physician, Jan Baptista van Helmont, writing at the beginning of the seventeenth century, even gave a recipe for eel manufacture: "Cut two pieces of grass sod wet with Maydew and place the grassy sides together, then put it into the rays of the spring Sun, and after a few hours you will find that a large number of small eels have been generated."

This claim, at least, would appear to be an innocent mistake. It can be traced to the tendency of observers, even astute ones, to lump together eels, snakes, and worms, especially when the creatures were small. But it is a little more puzzling that the distinguished German naturalist Athanasius Kircher (Figure 1.1), writing a few years later, reported that he had generated frogs from river mud. Kircher should have been clever enough to realize that the frogs were there all the time, surviving a dry period in the mud, and had not appeared spontaneously.

By the time that you are starting your studies at the beginning of the nineteenth century, belief in the spontaneous generation of such complicated animals has largely been abandoned by the world of science. The definitive experiments had been done almost a century and a half before your time, indeed not long after Kircher had announced his observations on the spontaneous generation of frogs. In 1668,

FIGURE 1.1   Athanasius Kircher
(1602–1680).

Francesco Redi (Figure 1.2), a Jesuit-trained doctor who had been appointed physician to the Medici family in Florence, published some
careful and thorough studies that for the first time applied the modern
scientific method to the question of spontaneous generation.

It was, of course, well known that when animals died, their bodies
were soon aswarm with tiny maggots and beetles. The beetles were obviously able to scuttle there from elsewhere, but the maggots were incapable of crawling for any great distance. Had these maggots been spontaneously generated? Redi addressed this question through the use of a
sensible scientific approach. He showed clearly that if meat was placed
in boxes and screened with muslin so that flies could not land on it,
maggots would not appear.

He left some boxes unscreened and found that in these boxes, maggots appeared in abundance—one of the earliest examples of a controlled experiment! He also demonstrated that the maggots were the
larvae of flies. They developed into a number of fly species, some of
which were new to him. He found that the same set of fly species appeared, regardless of the type of meat on which their larvae had fed.

Redi was inspired to do his experiments, not by any deep insight into
the nature of life, but rather by his belief that all living things had been

THE RISE AND FALL OF SPONTANEOUS GENERATION

FIGURE 1.2 Francesco
Redi (1626–1698).

created by God at the beginning of the world, and that "these worms
are all generated by insemination." And he was quite modest about his
results. He noted that in his own time, hunters and butchers protected
their meat by covering it with cloths, demonstrating that in this case
folk wisdom had outstripped science. Even Homer, he noted, seemed
quite aware of the propensity of flies to lay eggs on cadavers—in the
Iliad, Achilles worried that flies would lay eggs on the body of his friend
Patroclus while he was away avenging his death.

Redi's experiments, designed though they were to answer a narrow
question, did so clearly and unequivocally. With impeccable honesty,
however, he admitted that spontaneous generation might still be taking
place in other organisms. He tried to replicate his contemporary
Kircher's frog-generating experiment, and when he failed, modestly
suggested that he might simply have gotten the conditions wrong. And
he tried repeatedly, but without success, to determine whether the
wasps that developed inside galls, the swellings of plant tissue, appeared
spontaneously or came from wasps that had laid their eggs on the
plants. (The gall-wasp life cycle was worked out a few years later.)

As you learn about Redi's experiments, you are confident that he had
at least settled the maggot question. Nonetheless, the more general

question of spontaneous generation has continued to persist to your time. Some of the confusion that you are confronting in your studies stems not from conflicting scientific results, but from an ongoing battle between religious and secular forces.

Like Redi, the Catholic Church is opposed to spontaneous generation. The book of Genesis states that God had made living organisms only during the days of creation. Thus, if any living things are still appearing spontaneously, this contradicts statements in the Bible.

You also realize that during the Age of Enlightenment, many people revolted against unthinking acceptance of the Church's teachings. There was a tendency to equate belief in spontaneous generation with progressive, daring, and materialistic thinking. Many French philosophers and encyclopedists, under the leadership of Denis Diderot, embraced the idea of spontaneous generation, as did that iconoclastic naturalist the Comte de Buffon.

Even at the time of your studies, there are still plenty of cases of the sudden appearance of organisms that seem difficult to explain by any means other than spontaneous generation. One of these cases has emerged from the study of parasitology.

As dissection of dead bodies has become more common and accepted, pathologists are beginning to discover that the lungs and guts of people from all walks of life are full of a distressing assortment of roundworms, flukes, and tapeworms. Because the lungs and gut have openings to the outside, these worms might have entered the body from elsewhere, though even that possibility has been hotly disputed by the prestigious Buffon. Far harder to explain are the parasites buried deep in the liver, such as the hookworms and the leaflike liver flukes, with no obvious access to the world outside.

Are such parasites truly separate organisms with their own life histories, descended from a long line of ancestral hookworms and liver flukes? Did they infect Adam and Eve in the Garden of Eden? If so, then they must have been benign rather than harmful in those days, since everything else in the Garden of Eden presumably was! But if these parasites are an additional curse bestowed by God on Adam and Eve after the Fall, then God must have made them after the seven days of creation.

In your time, many scientists have escaped from such endless theological debates by assuming that liver flukes and hookworms appear spontaneously without the immediate intervention of God. It will be well into the nineteenth century before the complex life cycles of these parasites begin to be understood, finally putting to rest the possibility that they are spontaneously generated.

The World of the Very Small

As your studies advance, you find that some other remarkable observations about the living world can be interpreted as evidence both for and against spontaneous generation. The most striking of these is the discovery that the world is filled with tiny organisms, far smaller than the mice and frogs that Aristotle wrote about.

In the mid–1600s Anthony van Leeuwenhoek (Figure 1.3), a Dutch civil servant in Delft, began to make a series of simple microscopes. Each consisted of a single tiny but powerful lens, a kind of super-magnifying glass. A few of his instruments have survived, the best of which can magnify up to 280 times. It has been estimated that his finest microscopes, those that he reserved for his own use, could magnify an astonishing 400 times or more, quite enough to reveal bacteria.

By 1673 Leeuwenhoek was using his instruments everywhere to discover swarming worlds of hitherto unknown creatures. They seemed to be ubiquitous—in pond and lake water, in infusions of pepper, in cistern water that had been left standing, in the soil, in rotting meat, in his own semen. He recounted his discoveries in a series of long letters that he sent over a period of decades to the Royal Society in London. The scientific world was amazed.

Many people, including distinguished naturalists like the Chevalier de Lamarck, who worked a century later, were convinced that the most likely explanation for all these tiny organisms must be that they are arising by spontaneous generation. There are so many of them, and they seem to appear from nowhere! Clean water left standing soon acquires a whole bestiary of creatures. There must be some property of matter that makes them appear, some vital spirit with which the world is apparently infused and which can, under the right conditions, give rise to a wide variety of simple creatures.

Leeuwenhoek himself, however, did not fall into this trap. The beauty, complexity, and perfection of the creatures that he had discovered convinced him that they could not have arisen spontaneously from decaying matter. And he did not simply rest on this belief, but experimented with many of the tiny single-celled creatures that he had discovered. He observed that some of them, which we now call rotifers, could shrink down into an encysted form when the water in which they

FIGURE 1.3 Anthony van Leeuwenhoek
(1632–1723).

were living dried up (Figure 1.4). The cysts, he found, could be turned back into rotifers again by the simple addition of water.

Leeuwenhoek realized that the tiny cysts could easily blow away on the wind and be spread everywhere. In 1702 he wrote excitedly to the Royal Society about this discovery: " . . . [T]his most wonderful disposition of nature with regard to these animalcules for the preservation of their species . . . must surely convince all of the absurdity of those old opinions, that living creatures can be produced from corruption or putrefaction."

As you turn from Leeuwenhoek to scientists living closer to your own time, you find that many of his observations have been forgotten or misinterpreted. His statements and elegant demonstrations were sometimes unfortunately buried in his long, discursive letters to the Royal Society. As a result, the debate about spontaneous generation is by no means settled. It has continued down to your own time and has become fiercer than ever. Largely because of Leeuwenhoek's discoveries, the battleground has now shifted from visible animals such as worms and frogs to the world of the invisible.

THE RISE AND FALL OF SPONTANEOUS GENERATION

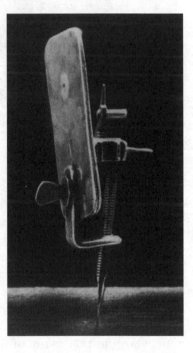

FIGURE 1.4   One of Leeuwenhoek's microscopes, and some of his drawings of rotifers.

The first shot of this new battle was fired in 1748 by an English lay priest, John Turberville Needham. He took hot mutton gravy fresh from the stove, sealed it in a flask, and found that it soon swarmed with bacteria. This, he claimed, clearly demonstrated the spontaneous generation of microbial life.

FIGURE 1.5   Lazzaro Spallanzani (1729–1799).

The opportunities for contamination in Needham's experiment were numerous, for among other things, he had not sterilized the flask beforehand. In 1765, Lazzaro Spallanzani (Figure 1.5), a distinguished Italian physiologist, published experiments refuting Needham's clumsy efforts. Spallanzani filled flasks with plant and egg infusions, sealed them, and set them in boiling water for an hour. They remained clear, whereas similar unboiled flasks soon became cloudy with swarms of microbes.

Needham immediately countered by claiming that Spallanzani had somehow destroyed some vital force in the liquids by boiling them for so long. Spallanzani's riposte was to boil his infusions for various times, then open them to the air. There was no sign that the most thoroughly boiled infusions had lost any vital force—indeed, the infusions that had been boiled for a long time tended to be thicker and more nutritious than those that had been boiled more briefly, and they soon swarmed with even more organisms.

Needham responded that the problem must lie with the air. In the boiled flasks, the elasticity of the air must have been damaged, preventing life from appearing. Spallanzani was unable to devise an experiment

FIGURE 1.6 Jean-Baptiste Monet, the Chevalier de Lamarck (1744-1829).

to disprove Needham's new criticism. Indeed, when he broke the seals on his flasks, he did hear a faint hissing—perhaps, he admitted, the air's elasticity had indeed been damaged.

You discover that this inconclusive battle has continued down to your own day. The results of these and many other experiments are now being mulled over by many people throughout Europe, with confusing and sometimes politically charged results.

On the practical side, Spallanzani's approach to sterilization has just been employed, in this year of 1810, with huge success by Nicolas Appert. He has demonstrated that it is possible to preserve meats and vegetables indefinitely in tin cans by sealing the cans and boiling them. His discovery is immediately put to use by Napoleon's war machine.

One might think that Appert's successful application of Spallanzani's approach should provide powerful evidence against spontaneous generation. But it is just at this time that Jean-Baptiste Monet, the Chevalier de Lamarck (Figure 1.6), is blithely ignoring these findings as he tries to fit spontaneous generation into a general theory of evolution.

Lamarck concludes that worms and protozoa are indeed still appearing spontaneously, but that they don't stay simple organisms for long. Driven by their felt needs *(sentiments intérieurs)*, they soon acquire sophisticated characteristics and pass them on to their descendants. In this way they will evolve into more complex organisms and leave their simple way of life behind. Because of spontaneous generation, the ecological niches in the living world that they have abandoned will soon be filled again by new, simple organisms that continue to appear through spontaneous generation. In Lamarck's view of the world, life is on a kind of endless evolutionary escalator.

Even Lamarck's contemporaries think that his theory is wildly wrong. But at least he is doing his best to try to explain evolution as an ongoing, continuous process, and in this he anticipates Darwin. At the same time, however, he is setting the stage for a vituperative debate, particularly in France, about creation versus evolution. The debate pits materialists, who use evidence for spontaneous generation to show that God need play no role in the origin of life, against members of orthodox religion, who are horrified at the very idea of spontaneous generation because it contradicts Scripture. It is religion and politics, not science, that drives this debate.

Probing the Nature of Life

This was the status of the origin of life during the first decade of the nineteenth century: unsettled and confused and riven by endless arguments. It is no wonder that, had you lived at that time, you would have been hopelessly puzzled.

As subsequent decades went by, a variety of complex and contradictory debates about the nature of life took center stage in the scientific community. Nothing, some scientists contended, could be more different than living and nonliving matter. As microscopists examined the contents of cells such as amoebas that are capable of movement, they noted that this material—termed *protoplasm* by the botanist Hugo von Mohl in 1846—had a complex structure and a remarkable ability to turn from a liquid to a solid and back. These observations led them to agree with Leeuwenhoek's original contention. Surely such complex material, so unlike anything to be found in the nonliving world, could only have arisen from previously living animals.

The apparent uniqueness of life led some to suggest that there was something mystical about it, some ineffable force that set it off from the nonliving world. This concept pervaded early nineteenth-century science. It was named *vitalism* by Georg Ernst Stahl in the eighteenth century and was expounded most eloquently in the nineteenth century by the physician Rudolf Virchow.

Virchow was a fierce partisan of progressive politics. He was once challenged by Otto von Bismarck to a duel, which he sensibly declined—even though one cannot help but feel a pang of regret at the lost opportunity. What might have happened to history had Virchow shot the Iron Chancellor?

He was equally fierce in his belief that all life had to arise from preexisting life. The organic and inorganic worlds were strictly separate. But even as he was defending vitalism, chemistry was making great strides and seemed to be showing just the opposite. As chemists began to examine the compounds found in the natural world and to compare them to those that they could make in the laboratory, they soon found that there was no obvious difference between them. The first real breach in the apparent wall between the living and nonliving worlds was made in 1828 by Friedrich Wohler, who used an inorganic salt, ammonium cyanate, to synthesize the simple organic compound urea, a compound found in the urine of mammals.

Some chemicals made by living organisms, the amino acids, turned out to be remarkably versatile, small organic molecules with both acidic and basic properties (Figure 1.7). The first amino acid to be isolated was asparagine, obtained from asparagus juice in 1806 by Nicolas-Louis Vauquelin and P. J. Robiquet. Other chemists, however, soon found that some amino acids present in living organisms could also be made in the laboratory. In 1850, Adolph Strecker synthesized an amino acid for the first time, using a mixture of acetaldehyde (derived from acetic acid, the major component of vinegar), hydrogen cyanide (a deadly poison), and ammonia. He called the amino acid alanine. Later, Thedor Weyl isolated this amino acid from silk. Strecker's synthesis, as we will see, prefigured in a remarkable way the prebiotic syntheses of amino acids conducted by Stanley Miller a century later.

The gap between the organic and inorganic worlds was apparently closing quickly. Was there really nothing that set life apart from the inorganic world? By the end of the nineteenth century, nearly all the amino acids commonly present in living organisms had been discovered. There are surprisingly few of them, and it was established that only twenty are found in all organisms. Even though various plants and animals do have some additional unusual amino acids, the twenty are

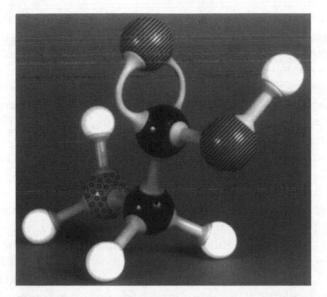

FIGURE 1.7   A molecular model of glycine (chemical
formula CH₂NH₂COOH), one of twenty different
amino acids found in the proteins of living organisms.
The black balls represent carbon atoms, the striped
balls oxygen atoms, the speckled ball a nitrogen atom
and the white balls hydrogen atoms. The other protein
amino acids have the same basic structure, but have
more carbon atoms and functional groups in side
chains that are attached to the bottom carbon atom.

the ones that appeared to be universal. Amino acids were also found to
be amazingly abundant in living organisms, and it was becoming apparent that they must make up a large fraction of living cells.

But at the same time, other chemists were showing that living organisms were more than just a collection of amino acids. In 1838, the
chemist Gerardus Johannes Mulder reported on some huge molecules
that he had isolated from milk and egg albumin. These molecules all
seemed to him to be much the same, since when he burned them he
found that they were all made up of large numbers of carbon, nitrogen,
oxygen, and hydrogen atoms, with traces of sulfur and phosphorus. Realizing their importance, he gave them the name proteins, after the
Greek word *proteios*, meaning "primary."

Before long it was realized that Mulder's proteins, at first thought to
be enormously complex molecules that could not be subdivided, were

really nothing more than collections of smaller amino acid molecules that had somehow been joined together by the activities of living cells. As the century advanced, the scientific community discovered that proteins are not the only large molecules found in quantity in living cells. The Swiss chemist Johann Friedrich Miescher (Figure 1.8), working in Germany in 1869, was the first to isolate a new class of very large acidic molecules. He obtained them from a disgusting source that only a truly dedicated scientist could have exploited: pus cells scraped from discarded surgical bandages. We now know that these molecules are found primarily in the nucleus of the cell, and they have been given the name *nucleic acids.*

Miescher found that the heads of salmon sperm were also a rich source of these acids, and he devised methods for purifying them for further study. He was able to show that, like proteins, these nucleic acids were made up of simple building blocks. Proteins are made up entirely of amino acids, but nucleic acids are built from three types of small molecules: sugars, phosphoric acid, and basic compounds that were later identified as purines and pyrimidines and are generically called nucleobases, or simply bases.

FIGURE 1.8　Friedrich Miescher (1844–1895).

THE SPARK OF LIFE

Long after Miescher's untimely death in 1895, it was discovered that there are two types of nucleic acids in all organisms. These two types, RNA (ribonucleic acid) and DNA (deoxyribonucleic acid), are named for small differences in their sugars—ribose and deoxyribose, respectively. DNA, as everybody now knows, resides in the nucleus of the cell and carries the cell's genetic information. This information, encoded by the sequence of bases in the DNA, is transported by RNA from the nucleus into the surrounding cytoplasm of the cell. There it is used to specify the construction of proteins.

So both sides in this controversy about the nature of life turned out to be partly right. Living organisms contain complicated molecules, which makes them unique, but these molecules are in turn made of much simpler building blocks. Even though proteins and nucleic acids somehow confer remarkable properties on living organisms, they are nonetheless made up of these simple blocks. The mystical idea of vitalism eventually vanished from the laboratories of chemists, although the concept of a mysterious life force persisted into the twentieth century in the form of the *élan vital* of the philosopher Henri Bergson.

The Universe Is Dissymmetric!

Was life nothing but chemistry after all? The argument seesawed back and forth throughout the early part of the nineteenth century. It soon became apparent that even if living organisms were simply collections of chemicals, they were very clever collections indeed.

Starting in 1812, the French physicist Jean Baptiste Biot (Figure 1.9) made a series of remarkable discoveries, even though they at first seemed to have nothing to do with living organisms. When a beam of light is reflected from a mirror, it becomes polarized—the light waves oscillate only in one plane. Biot shone polarized light through quartz crystals, and found that the crystals occurred in two forms, identical in appearance but able to rotate the plane of the light in equal but opposite directions. He named the two forms left- and right-handed quartz and characterized each form as having optical activity.

He soon found that optical activity also applied to some compounds obtained from organic material. In careful studies, he showed that solutions of sugar, camphor, and turpentine also rotated polarized light. But each of these different organic solutions exhibited a characteristic type of optical activity. Whatever their source, unlike the quartz crystals, they

FIGURE 1.9   Jean Baptiste Biot (1774–1862).

were all made up of only one form and rotated the plane of the light in only one direction.

Biot suggested that both the quartz crystals and the organic solutions had some sort of asymmetric property that contributed to the observed optical rotation. Quartz, he reasoned, must consist of two types of crystals that are mirror images of each other. The first tiny seed crystal that forms determines whether a left- or right-handed crystal will be produced. Because each type of seed crystal can form with equal probability, equal numbers of the two types of crystals are produced. As a result, quartz on Earth (and, we now know, elsewhere in the Solar System) consists of a fifty-fifty mixture of left- and right-handed crystals.

Organic molecules, on the other hand, are different. Sugar solutions, for example, rotate the plane of light to the right. Living organisms do not seem to make any sugars that rotate the plane of light in the other direction. Biot realized that there truly is a remarkable difference between the substances found in the living and nonliving worlds. Living organisms seem to make only one type of optically active molecule, whereas in the nonliving world—at least in the world of quartz crystals—there is no such capacity.

But perhaps crystals and solutions are somehow different, and the difference between them has nothing to do with asymmetrical molecules. In 1850, the young French chemist Louis Pasteur (Figure 1.10) ruled out this possibility, in experiments that launched his brilliant scientific career. While growing tiny crystals of the compound tartaric acid that he had made in the laboratory, he observed that the crystals were of two mirror-image shapes. Like the quartz crystals of Biot, they could actually be distinguished simply by observation.

The crystals were far too small for Pasteur to pass beams of polarized light through them. So he sorted out the tiny left- and right-handed crystals of the acid into two piles using a magnifying glass and then dissolved each separated pile in water.

As he had anticipated, the two solutions rotated beams of polarized light in equal but opposite directions. The directions of rotation correlated with the mirror-image shapes of the crystals from which the solutions were made. A solution made up of equal proportions of the two

FIGURE 1.10   Young Louis Pasteur (1822–1895).

THE RISE AND FALL OF SPONTANEOUS GENERATION

crystals (now known as a *racemic mixture*) produced no rotation. Pasteur had demonstrated that both Biot's crystals and optically active solutions were really examples of the same underlying phenomenon of molecular asymmetry.

Biot, by this time an elder statesman of French science, was uncertain what to make of Pasteur's bold claim. Before he would agree to present the results to the French Academy of Sciences, he insisted that Pasteur repeat the experiments before his very eyes. Providing Pasteur with tartaric acid crystals he had prepared himself, Biot watched as Pasteur carefully separated the crystals into the left- and right-handed piles. Then Biot asked Pasteur to leave the room as he prepared solutions of the crystals to examine with polarized light. When Biot was ready to begin the measurements, Pasteur was invited to return.

As Pasteur recalled later in his life, Biot "was very visibly affected" when he saw the clear difference between the solutions. Grabbing Pasteur's hand, Biot cried out, "My dear child, I have all my life so loved the sciences that this makes my heart throb with joy."

A year later, Pasteur discovered that there could be similar dramatic differences between molecules purified from living organisms and those synthesized in the laboratory. He prepared solutions of asparagine from asparagus plants. When polarized light was shone through them, the plane of the light was rotated to the left. But synthetic asparagine that he had made in the laboratory showed no rotation at all, and he realized that this was like his mixture of equal amounts of left- and right-handed tartaric acid crystals. The synthetic asparagine must be a racemic mixture.

Pasteur concluded that living organisms must somehow carry out the same kind of process that he had accomplished with tartaric acid. But it was obvious that they did not use a magnifying lens and tweezers! In order to accomplish his separation, he had been forced to wait until the crystals grew, but living organisms could somehow carry out this amazing process at the molecular level. Such feats were quite beyond the capabilities of the chemists of Pasteur's day.

So life did turn out to be different from the organic chemistry of the laboratory after all—not in the building blocks of life, which are quite common and simple molecules, but in the way in which they are made and then linked together. Living cells are able to make huge and dauntingly complex protein and nucleic acid molecules that the biochemists of the time were quite powerless to synthesize in the laboratory. And even the building blocks of proteins are different—they are made up of amino acids of only one optical type, a feat that also could not be accomplished in the bubbling retorts and condensers of an organic chemistry laboratory.

How could living cells achieve these results? Pasteur felt that some sort of dissymmetric force or action had to be responsible. Perhaps the force had something to do with magnetism. He crystallized tartaric acid in the presence of a strong magnetic field, but failed to find any enrichment in the amounts of either left-handed or right-handed crystals. In the end, Pasteur was left to exclaim that "l'univers est dissymétrique." Remarkably, Pasteur's speculation anticipated the discovery made a century later in the 1950s that the electrons given off during the beta decay of radioactive elements are asymmetric and have a preferred handedness.

The Battle Between Pasteur and Pouchet

In the meantime, the battle over spontaneous generation continued apace. In 1837, Theodor Schwann finally succeeded in putting to rest the old argument about the elasticity of the air. After heating air strongly to kill any floating organisms, he introduced it into a flask of sterile broth, and found that the broth remained sterile. But proponents of spontaneous generation, infinitely resourceful at retreating to fallback positions, responded by claiming that when Schwann heated the air, he damaged it. They performed further experiments that seemed to contradict those of Spallanzani and Schwann. It seemed that certain liquids, particularly infusions made from dried grasses, could not have their vital force destroyed by boiling. Even after these hay infusions had been boiled in sealed flasks, they quickly swarmed with tiny organisms. Some kinds of organic liquids are able to generate microbes even in the presence of "damaged air."

The climactic period of debate began in 1859, with the publication of a long monograph by Félix Pouchet (Figure 1.11), the director of Rouen's Natural History Museum. Pouchet claimed to show, using hay infusions, that spontaneous generation can occur even when sterile air is passed through mercury and then introduced into the flask.

It was at this point that Louis Pasteur joined the battle, against the advice of the esteemed Biot and Jean Baptiste Dumas, the most prominent French chemist at the time. Pasteur was driven to do so because he was in the middle of an important series of experiments on fermentation. He was sure that different fermentations, such as those that

FIGURE 1.11   Félix Pouchet (1800–1872).

produce alcohol and vinegar, could be traced to the activities of specific microorganisms. But he realized that his results could not be correct if microbes were arising by spontaneous generation. If a great variety of microbes could appear in any liquid at any time, then fermentation should always produce a mixture of products. Because this did not happen, the idea of spontaneous generation must be wrong.

Beginning in 1860, he published a devastating series of experiments that appeared to demolish Pouchet. Using sterilized infusions of boiled yeast and sugar water, Pasteur attacked the problem in many ways.

The first series of experiments began with a bit of asbestos that had been used as a filter to trap particles in the air that Pasteur suspected were the source of microbial contamination. When the asbestos was dropped into a sterile solution, it immediately generated growth. But if it was first heated so that any trapped organisms were killed, no growth took place. To counter Pouchet further, Pasteur showed that even mercury can contain germs unless it is sterilized first: The mercury through which Pouchet bubbled his air must itself have been contaminated.

More elegantly, Pasteur pulled out the necks of flasks that contained infusions of various organic substances into a variety of sinuous swan shapes (Figure 1.12). He then broke off the tips of the extended necks and boiled some of the flasks for a few minutes while leaving several unboiled for controls. The unboiled flasks became infected with microorganisms within a day or so. The boiled flasks, however, remained sterile indefinitely even though air was quite free to pass in and out. As air diffused down the long necks, any germ-carrying dust particles adhered to the sides of the glass tube before they had a chance to reach the liquid in the body of the flask. If the boiled flasks were tipped so that the liquid inside came in contact with the outer parts of the long neck, microorganisms soon flourished in the flask. At last, after a hundred years of argument, these experiments finally laid to rest the old claim repeatedly raised by believers in spontaneous generation: that somehow the vital force in the air is damaged by the process of sterilization.

Pouchet and the other proponents of spontaneous generation, replied by pouring scorn on the idea that the air is full of living organisms. Surely, if Pasteur's view of the world was correct, there should be so many of these little creatures floating about that we should all be living in a fog of organisms! Pasteur's elegant riposte was to sterilize large numbers of flasks, carry them to different altitudes, then briefly break

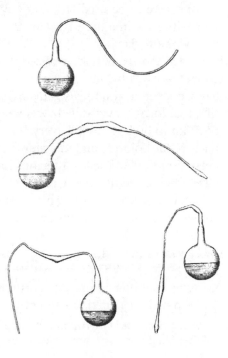

FIGURE 1.12 Pasteur's flasks with their drawn-out, curved necks. From *Oeuvres de Pasteur*, edited by Pasteur Vallery-Radot (Paris: Masson et Cie, 1922), 2:260.

THE RISE AND FALL OF SPONTANEOUS GENERATION

open their necks and seal them off again. Flasks opened in Paris almost all became contaminated, whereas most of those opened on the glacier at Montenvers, two thousand meters above sea level, remained sterile. Flasks opened at intermediate altitudes became contaminated about half the time.

These experiments demonstrated that small volumes of clean air at high altitude could indeed be sterile. Pasteur pounded home the point that it is the floating matter in the air that is the source of the contamination. The filthy air of Paris, filled with dried fragments of horse dung and droplets from innumerable sneezing and coughing humans, was a far more potent source of contamination than the clean air of the mountains.

Pouchet immediately repeated Pasteur's experiment using his own hay infusions and found that all of them became contaminated regardless of where they were opened. This set the stage for the final confrontation between Pasteur and Pouchet. In 1864 the French Academy of Sciences announced a contest in which the altitude experiment was to be repeated. Pasteur's adherents in the academy set up the conditions for the contest. Pouchet and his collaborators agreed at first to participate, but when they discovered that only flasks filled with Pasteur's infusions were to be used and that hay infusions would not be allowed, they withdrew in a huff. Pasteur, triumphant on the field, appeared to have won the day.

In fact, though nobody realized it at the time, Pouchet's results were correct—it was his interpretation of them that was in error. Wisps of hay are covered with tiny organisms, such as the bacterium *Bacillus subtilis*, that form heavy-walled cysts when desiccated. These cysts are very resistant to boiling. To kill everything in a hay infusion, you must raise it above the boiling point of water for an extended period. Had Pouchet been permitted to use his infusions in the competition, he would surely have obtained results that were different from those of Pasteur.

Pasteur had been pursuing his studies with the utmost rigor in order to answer immediate questions such as the role of living organisms in fermentation. But he and his conservative supporters in the academy were well aware of the wider implications of the studies. In a talk given in 1864, he set out clearly the way in which his experiments disproving spontaneous generation reinforced the orthodox religious view of the world and undermined the materialist and evolutionary heresy of Lamarck:

> What a victory for materialism if it could be affirmed that it rests on the established fact that matter organizes itself, takes on life itself. . . . Thus, admit the doc-

trine of spontaneous generation and the history of creation and the origin of the organic world is no more difficult than this. Take a drop of sea water containing some nitrogenous material . . . and in the midst of it the first beings of creation take birth spontaneously. Little by little they transform themselves . . . for example, to insects after 10,000 years and to monkeys and man after 100,000 years. Do you now understand the link which exists between the question of spontaneous generation and those great problems I listed at the outset?

It is ironic that Pasteur's superb experiments, which eventually led to the germ theory of disease and to immense benefit for humankind, should have been embraced so wholeheartedly by religious and political conservatives. It is rather sad in retrospect that Pasteur encouraged this interpretation.

The Battle of Tyndall's Boxes

Pasteur seemed to have won the debate, but the proponents of spontaneous generation had not given up. The final encounter took place late in the nineteenth century.

This final set of experiments hinged again on the question of floating matter in the air. Elegant though Pasteur's demonstrations were, they always depended ultimately on inference—he could not see the germs on the dust particles that filled the air, even though his experiments told him that they must be there.

But the particles themselves, at least, could be seen. The English physicist John Tyndall (Figure 1.13) gained his reputation by showing that the blue color of the sky is due to the scattering of light by tiny particles of dust and water droplets. From this observation, he was led to an investigation of the behavior of these particles.

He began by exploring the well-known phenomenon that the tiniest speck of dust in the air can be illuminated against a dark background if a powerful beam of light is shone at right angles to the observer. Using this technique, he began to look at how much floating matter there was in different kinds of air.

One of his most intriguing discoveries was inspired by some observations made by the surgeon Joseph Lister, who was himself a strong proponent of Pasteur's ideas and who pioneered sterile techniques in surgical procedures. Lister often saw patients who had suffered blows to the chest that broke their ribs, piercing the lungs but leaving the skin intact. He was struck by the fact that as long as the damage was all internal, there was no infection, even though the patients were of course

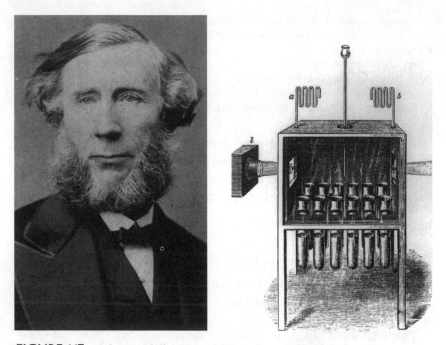

FIGURE 1.13   John Tyndall (1820–1893) and one of his boxes.

breathing nonsterile air that must, as it passed into the lungs, be reaching the site of the wound.

Tyndall suspected that the lungs, with their innumerable mucus-lined passages, must trap and remove any infective particles from the air before they could reach the wound. He confirmed this by blowing air from his own lungs into his illuminated chamber. The air from deep in his lungs contained no floating specks at all—it was utterly transparent.

He was soon attacked for these ideas by a persistent critic named H. Charlton Bastian, who played Pouchet to Tyndall's Pasteur. Irritated by the attack, Tyndall concentrated his attention on the final solution to the spontaneous-generation problem.

By 1876, he had managed to replicate the sterilizing effects of lungs in the laboratory, employing a simple device that became known as a Tyndall box (Figure 1.13). He coated the insides of a black-painted box with a thin layer of sticky glycerin. When the box was put in a quiet place, all the floating particles inside it soon settled out or collided with the sides and became trapped. The air inside the box could be seen to be completely transparent when a beam of light was shone through it.

At this point, sterile solutions of any sort that were exposed inside the box would remain sterile indefinitely. Tyndall boxes were put on display at the Royal Society in London, where they convinced everyone who saw them. Although Bastian insisted for many years that spontaneous generation was a reality, he had become a quaint voice from the past.

The battle of Tyndall's boxes was the last skirmish in a war that had lasted three hundred years, from the time of Francesco Redi and Anthony van Leeuwenhoek to the last desperate shots fired by Bastian. It had consumed many lifetimes and had been fought at the political and philosophical as well as the scientific levels. By quite literally settling the dust of this final skirmish, Tyndall managed to resolve the matter. He also closed the great circle of inference on which the new science of epidemiology depended: that diseases caused by microbes could be spread invisibly through the air. It was an inference that had begun with Leeuwenhoek's observation that dried rotifer cysts can blow away in the wind, but that even the brilliant work of Pasteur had not quite managed to prove irrefutably.

There was now no doubt that at least under the conditions found in the familiar, present-day world, life did not appear spontaneously. Spontaneous generation was down, though as we will see later, it was not totally out. But the next fight, the one that Pasteur had foreseen so clearly, was only just beginning. The great question of where and when life had arisen seemed more puzzling than ever.

Flash! Kelvin Traps Darwin in a Prison of Time!

During the period that the idea of spontaneous generation was undergoing an endlessly prolonged demise, a new and vibrant idea about the process of evolution was displacing the clumsy, spontaneous-generation-based ideas of Lamarck. Charles Darwin's *Origin of Species*, when it appeared in 1859, suddenly illuminated the world of living things in a blaze of light (Figure 1.14). His mechanism of natural selection provided a means by which simple organisms could have evolved into more complex ones. But although Darwin's insight could explain the immense diversity of the living world that we see around us, he was quite unable to probe the beginnings of life except to say that it seemed to have begun only once—and even for this conjecture, he had no proof.

If life had evolved by Darwin's mechanism from simple to complex organisms, this must have required enormous amounts of time. As he

FIGURE 1.14 Caricature of Charles Darwin (1809–1882) that appeared in *Vanity Fair*, September 30, 1871.

visualized it, evolution is a slow process that takes many thousands of generations—particularly because his theory assumes that major changes must be the result of the summation of many small, selectable changes. But the huge spans of time required for evolution, as he was soon to be brutally shown, did not appear to be available.

In the mid-nineteenth century, the leading authority on the age of the Sun and the Earth was William Thomson, a professor of physics at the University of Glasgow, later to become Lord Kelvin (Figure 1.15). Around the time that Darwin published his *Origin*, Thomson was estimating the age of the Sun as a mere 20 million years. He calculated that even if the Sun were made of solid coal, it could not have burned for any longer without running out of fuel.

Of course, this meant that the Earth had to be the same age or even younger. And the Earth's age, too, could be estimated using basic nineteenth-century physics. Thomson assumed that the Earth had started out as a globe of molten rock when the Solar System formed, and that as it aged, heat was lost to space. Because all the initial heat had not been lost by Thomson's time, then the Earth's interior must still be hot.

To find out just how hot, he measured the temperature in deep coal mines and quantified what miners had known for centuries, that the

FIGURE 1.15   William Thomson (Lord Kelvin, 1824–1907) in 1846 when, at the age of twenty-two, he was appointed Professor of Natural Philosophy at the University of Glasgow. His inaugural essay (in Latin!) dealt with the heat distribution within the Earth and what this implied about its age, a subject that would concern him for the rest of his life. From A. Gray, *Lord Kelvin: An Account of His Scientific Life and Work* (London: J. M. Dent & Co., 1908).

temperature of the Earth increases with depth below the surface. Because at the time there was no other known source for the Earth's internal heat, Thomson used this temperature gradient to calculate how long it had taken the Earth to cool to its present state. He estimated that the Earth was between 20 and 40 million years old, an age range in approximate agreement with his earlier estimate of the age of the Sun. Later, the American geologist Clarence King refined Thomson's approach and arrived at a value of 24 million years.

Darwin was devastated by the force of this argument. If the Earth was only 24 million years old, this simply did not give enough time for modern complex and multicellular organisms to have evolved from a simple, single type of primitive organism by the small steps permitted by natural selection. To make things worse, it would have taken much of that 24 million years for the Earth to cool from a molten state to one that was habitable by living organisms!

Darwin had allies, however, in the geological community. Many geologists were uncomfortable with the estimates of Thomson and King, because they knew from rates of erosion and sedimentation that geological features such as river valleys must have taken a long time to form. Indeed, the reconstruction of the slow processes by which most of the Earth's surface must have formed, by Charles Lyell and others, had led to the foundation of an entire school of geology. The credo of this "uniformitarian" school, based on the supposition that vast periods of time had passed during the Earth's history, was summarized by the

THE RISE AND FALL OF SPONTANEOUS GENERATION

late eighteenth-century geologist James Hutton: "No vestige of a beginning, no prospect of an end."

Unfortunately, this uniformitarian argument was a rather vague and hand-waving one. Once-mighty mountains had obviously been worn down over long spans of time, but just how long this had taken could not be determined with the precision of Thomson's temperature measurements. Less vague clues to the antiquity of the Earth could be found, although they too were based on flawed assumptions.

For example, in 1899 the Irish geologist John Joly presented a paper to the Royal Dublin Society in which he calculated the age of the Earth by using the fact that river water contains a small amount of the element sodium. He showed that as a result of the slow accumulation of sodium and other elements from this source, the oceans should have reached their present level of saltiness (salinity) in 89.2 million years.

We now know that the Earth is so old that if all the salt in the oceans has come from rivers, as Joly supposed, seawater would be saturated with salt by this time. It turns out that Joly had neglected the role of mineral deposition in removing sodium from the ocean, along with the activities of certain marine organisms that make their exoskeletons from some of the other constituents of seawater. These processes have kept constant the salinity of the ocean, which indeed was approximately as salty in ancient times as it is today. Nonetheless, Joly's calculation was a legitimate estimate for the Earth's minimum age—it certainly could not be any younger than about 90 million years.

Thomson was unconvinced of the validity of Joly's age estimate (after all, it was based on the imprecise sciences of chemistry and geology, not physics!), as well as those made by the geologists of the uniformitarian school. Because of his stature as the leading physicist of the time, most of the scientific community sided with him and considered all the other estimates to be less creditable.

Panspermia Brings Temporary Relief

Thomson had brought the question of the origin of life on Earth to an impasse. If spontaneous generation was not happening, and the Earth was too young to be compatible with Darwinian evolution, where could life have come from? The possibility that life could have originated in the first place from inorganic matter, much less reached its present complexity during the few tens of millions of years of the Earth's brief history, seemed remote and improbable. The time was ripe for the re-

FIGURE 1.16
Svante Arrhenius (1859–1927) in 1904.
(Courtesy Gustaf Arrhenius)

vival of an old idea: Life on Earth might have originated from a seed or spore arriving from some other world.

Although the possibility that life had come from elsewhere can be traced to the early part of the nineteenth century, it was the German physician Hans Eberhard Richter who broadened it into a "cosmozoa" concept in 1865: "We regard the existence of organic life in the Universe as eternal. Life has always been there; it has always propagated itself in the shape of living organisms, from cells and from individuals composed of cells." By 1880, Wilhelm Preyer had developed the eternal life idea in detail, using Pasteur's disproof of spontaneous generation to conclude that life must always have existed throughout the universe. In other words, life was eternal, just like matter.

John Tyndall had given a name to the idea that life could float everywhere in the atmosphere on tiny motes of dust. He called it *panspermia*. The original meaning of this term was soon abandoned. Instead it was extended to encompass the idea that life could float freely through space, permeating the universe. This new panspermia theory was rapidly adopted, although in somewhat differing forms, by many scientific giants of the time, including Thomson, the German physiologist Hermann von Helmhotz, and the Swedish chemist Svante Arrhenius (Figure 1.16).

Thomson was one of the more vocal proponents. In a lecture in Edinburgh in 1871, he made a special plea for the idea: "The ideas that life originated on this Earth through moss grown fragments from the ruins

THE RISE AND FALL OF SPONTANEOUS GENERATION

of another world may seem wild and visionary; all I maintain is that is not unscientific." Thomson was the subject of much ridicule in the popular press for embracing this idea, such as this satirical poem that appeared in *Punch* in August 1871:

No Conjuror's Conjecture
Could a meteoric stone,
Pray, Sir William Thomson,
Fall, with lichen overgrown?
Say, Sir William Thomson,
From its orbit having shot,
Would it, coming down red-hot,
Have all life burnt off it not?
Eh, Sir William Thomson?
Not? Then showers of fish and frogs
Too, Sir William Thomson,
Fall; it might rain cats and dogs.
Pooh, Sir William Thomson.
That they do come down we're told.
As for aerolite with mould,
That's at least too hot to hold
True, Sir William Thomson.

Despite the popular ridicule, Thomson had strong support throughout the scientific community, and several noted scientists came to his defense. Indeed, the panspermia theory at first appeared to be supported by evidence from meteorites. In 1880, a geologist named Otto Hahn (not the Otto Hahn of radiochemistry fame) published a book titled *Die Meteorite (chondrite) und ihre Organismen* (The chondrite meteorites and their organisms), in which he claimed that the meteorite Knyahinya, which fell in Hungary on June 6, 1866, contained signs of extraterrestrial sponges, corals, and plants.

The meteorite claims, at least, were soon challenged. Pasteur had looked for bacteria in several meteorites, including the Orgueil meteorite, which had fallen in France two years before the arrival of Knyahinya. He failed to find any evidence of life or indeed of any substances associated with life. He apparently thought that these observations were too trivial to publish, for they are found only in his notebooks.

Nonetheless, as more meteorites were investigated in chemical detail, a few were found to contain small amounts of carbon-based compounds. These meteorites with organic compounds were clearly different in composition from the more common chunks of nickel and iron that make up the majority of meteorites. Richter cited this carbon from outer space as support for the existence of life in far-off worlds.

Attacks on the meteorite-based panspermia theory soon came from many quarters. Long exposure to radiation in space would surely sterilize the meteorite. More dangers would have awaited the few radiation-resistant spores that might have survived. Even Arrhenius, a panspermia proponent, admitted that during atmospheric infall to the Earth, a meteorite would be broken up into small, heat-sterilized fragments. He also thought that any imaginable process by which a meteorite could have been blasted off the surface of some other planet in the first place would generate conditions far too extreme to permit the survival of any organisms.

Nonetheless, Arrhenius managed to introduce ingenious revisions to the panspermia theory, in order to overcome the problems of survival for long periods in outer space. Suppose, he suggested, that bacteria drifted as individual spores through space, rather than hitching a ride on meteorites. Because such spores are very small, on the order of a few thousandths of a millimeter in diameter, they could be transported throughout the universe by the pressure exerted by starlight. A spore wafted into the Earth's stratosphere might be pushed further outward by the Sun's radiation and could reach high velocities as it was lifted into space. He calculated that it would be expected to reach the outer limits of our solar system in only a few years, and that it could reach the nearby star Alpha Centauri in nine thousand years.

Because outer space is so cold (below −200°C at the orbit of Neptune), and there is no oxygen, the spores might be able to retain their germinating power during such a journey. As they reach the first wisps of the upper atmosphere of a planet like Earth, these almost weightless spores would slow and drift down gently without being heated.

Arrhenius's version of panspermia may seem far-fetched to most readers today. But at the turn of the twentieth century, very little was known about even our immediate neighbors in the Solar System. Indeed, there was every reason to suppose that life could exist on nearby planets. Venus, with its thick cloud cover, was imagined to be a verdant tropical paradise instead of the roasting desert that recent space probes have unkindly revealed. And it was during the early years of the twentieth century that the American astronomer Percival Lowell filled the newspapers with stories about channels for water on Mars that had been built by Martians constructing planetwide irrigation projects.

At the start of the twentieth century, panspermia was essentially the only game in town—it seemed to be the only possible explanation for the origin of life. And even today, there is something very beguiling about panspermia. Of course, nobody now believes that life is an inherent property of the universe—it must have originated after the formation of the universe itself.

But suppose that the likelihood that life will appear on a given planet is very low. Francis Crick, the codiscoverer of the structure of DNA, wrote a book in 1981 called *Life Itself*. In it, he suggested with a nearly straight face that intelligent organisms elsewhere in the universe might have seeded planets like Earth with life by scattering specially designed life-containing chunks of rock in all directions. Given that the appearance of life is sufficiently unlikely, an intelligent agent would be far more effective than accidental panspermia at spreading it from planet to planet.

Crick later backed away from his claim. He admitted that when he wrote the book he had not realized the ability of natural selection to speed up the process of evolution. Of course, he had also not considered the immense spans of time necessary for these rock-transported life-forms to travel from one end of the universe to the other, or how these hypothetical, intelligent creatures could have managed to obtain all the material for their enormous and wasteful seeding operation. Finally, and most importantly, where did these Johnny Appleseed creatures come from in the first place?

Doubtful claims of traces of life, and even of living organisms, in meteorites have persisted throughout the twentieth century. One highly controversial example was the report in 1932 by the bacteriologist Charles Lipman at the University of California at Berkeley, who claimed to have cultured live bacteria from carefully prepared samples from the interiors of several meteorites. This prompted a story in the *New York Times* with the headline "Are Meteorites Alive?" Subsequent studies by others could not confirm Lipman's results, and it was eventually concluded that the bacteria were nothing more than terrestrial contaminants that had penetrated the meteorites after they fell to the Earth. Unfortunately, the lessons learned from this tale were soon forgotten. As we will see later, the issue of life in meteorites is still with us today and the same problems of interpretation have arisen.

A proof or disproof of panspermia will have to wait until life on other planets is examined in detail. If the chemistry of these organisms is exactly the same as ours, Arrhenius's ideas could be vindicated. If it is different, then life might have arisen independently (though even such a dramatic finding would not rule out panspermia completely).

The panspermia idea simply pushed the problem back without solving it. Life had to begin somewhere, and the explanation proposed by Richter that life is somehow an integral property of the universe verged on the supernatural. Whether life began on Earth or on some other planet, it must have done so through a specific sequence of chemical events. Further, these events should eventually be repeatable in the laboratory. New thinking was desperately needed. It was time for the revival of spontaneous generation, not in the form originally proposed by Aristotle and the scholars who succeeded him, but in a new, more scientific, and—most excitingly—more testable form.

2

PRIMORDIAL SOUP

If (and oh! what a big if!) we could conceive in some warm little pond, with all sorts of ammonia and phosphoric salts, light, heat, electricity present, that a protein compound was chemically formed, ready to undergo still more complex changes. At the present day, such matter would be instantly devoured or absorbed, which would not have been the case before living creatures were formed.

Charles Darwin, writing to botanist Joseph Hooker in 1871

IN MOSCOW'S NOVODEVICHY CEMETERY, NEAR THE SOARING bell tower of Smolensk Cathedral, lie buried some of the major figures of Russian art and literature—and also some of the major figures associated with the country's experiment with Communism.

One of these is Nikita Khruschev. Another, whose grave is marked by a dramatic black stone column topped by a grim-visaged bust with an aggressive Lenin-like goatee, is Aleksandr Ivanovich Oparin (Figure 2.1). Ironically, Oparin lies buried in the cemetery of a monastery complex recently designated "the heart of Holy Russia." Russia also experimented officially with atheism during the time that Oparin did his scientific work, and briefly provided him with the intellectual climate that allowed his work to flourish.

In 1924, as a thirty-year-old biochemist, Oparin published a short book about a remarkable idea. He suggested that simple organic compounds similar to those associated with life had been synthesized by natural chemical processes on the early Earth. He theorized that as the molten Earth cooled, carbon that had been liberated from its bonds with metals would react with hydrogen to form hydrocarbons. He may have derived this idea from the work of an earlier Russian chemist, Dmitri Ivanovich Mendeleev, who is considered the father of the periodic table of elements and had proposed a similar scheme to explain the origin of petroleum.

FIGURE 2.1   Aleksandr Oparin (1894–1980) in his office at the Institute of Biochemistry in Moscow, circa mid-1970s. (Courtesy Antonio Lazcano)

Because ammonia would also have been present, produced by the combination of hydrogen with nitrogen, cyanide and therefore amino acids would soon have formed. Oparin reasoned that these simple chemicals interacted further with each other to produce more and more complex substances. As a consequence of this chemical evolution process, which was only vaguely defined, life somehow emerged from a primordial, prebiotic soup.

Oparin also made another important point. All these chemical reactions had, of course, happened in the absence of life, which would otherwise have quickly destroyed these evolving compounds. Here, without realizing it, he harked back to Darwin's reasoning: Once life has begun, it can never arise again de novo, because the newly arising organisms will be eaten by existing life.

In 1928, the thirty-six-year-old British geneticist and scientific polymath John Burdon Sanderson Haldane published a similar idea quite independently (Figure 2.2). Like Oparin, Haldane noted that soon after the early Earth had cooled and a solid surface formed, there would have been no oxygen present in the atmosphere. Although the present atmo-

FIGURE 2.2 J. B. S. Haldane (1892–1964) addressing a United Front rally in Trafalgar Square, London, 1937. From R. Clark, *The Life and Work of J. B. S. Haldane* (London: Hodder and Stoughton, 1968).

sphere contains 21 percent oxygen (the rest consists of 78 percent nitrogen and 1 percent argon), the oxygen is a product of photosynthesis and thus only accumulated in the atmosphere after this biological process arose. Haldane suggested that the early atmosphere might have been made up primarily of carbon dioxide and ammonia.

Haldane knew that the Earth's surface is today protected from solar ultraviolet light, a protection actually generated by the ultraviolet light itself. This energetic form of light acts on free oxygen that diffuses up into the stratosphere from the atmosphere below, producing a layer of ozone. The ozone in turn acts as a barrier to ultraviolet and prevents it from penetrating to the surface of the Earth. But without oxygen in the atmosphere, an ozone layer would not have been present and the early Earth's surface would have been bathed in ultraviolet radiation. Haldane did not imagine, of course, that one day human activities would damage the ozone layer, resulting in the dangerous thinning of ozone over the polar regions in the latter part of the twentieth century.

Because the early atmosphere would have been transparent to ultra-violet light, this radiation would have been a powerful energy source, resulting in further chemical reactions. Haldane suggested that a dramatic kind of chemistry might therefore have taken place on the early Earth:

> [When] ultra-violet light acts on a mixture of water, carbon dioxide and ammonia, a vast variety of organic substances are made, including sugars and apparently some of the materials from which proteins are built up. . . . [B]efore the origin of life they must have accumulated till the primitive oceans reached the consistency of hot dilute soup.

In spite of these authoritative-sounding statements, Haldane had no direct evidence that this could happen! He was assuming that this is the way chemistry ought to work.

During the rest of his long and active scientific career, Haldane continued to publish articles that contained speculations about the origin of life, the "hot, dilute soup," space travel, cosmology, and life in the universe. He even returned to panspermia, suggesting the existence of "astroplankton," which could seed planets on which conditions suitable for life existed. In short, he was not wedded to one particular idea—his mind exploded across its own universe of possibilities.

Haldane, like H. G. Wells, apparently had the happy faculty of foreseeing the future. By estimating from chemical principles how long it would have taken for life to have arisen spontaneously in the primitive oceans of the Earth, he concluded that life had appeared suddenly—a conclusion reached in the 1990s by a number of scientists working in the origin-of-life field. And in 1964, near the end of his life, he also suggested that ribonucleic acid (RNA) may have played an even more important role in the first living organisms than it does today. This idea was also well ahead of its time, for only during the 1980s and 1990s has evidence appeared that supports it.

Adding Ingredients to the Warm Little Pond

The ideas of Oparin and Haldane were remarkable quantum leaps in our understanding of how life began on Earth. They were the first scientists to wrestle with the real chemical questions. The proponents of the panspermia theory had never done this—they had merely side-stepped the question by pushing life's origins back to a time long ago in some solar system far, far away.

Oparin further refined his ideas, first in the book *Origin of Life*, published in Russian in 1936 (see Plate 1) and then in a 1957 book, *The Origin of Life on the Earth*. Both books were swiftly translated into English and widely read. Like Haldane, he reasoned that the first living organism must have lived off the fat of the land in the form of the prebiotic soup.

The 1936 and 1957 books presented many detailed possible reaction pathways leading to complex molecules and were backed up by numerous references to published papers. Even today, these books remain valuable resources for information about the synthetic methods invented by many different chemists during the early years of organic chemistry. This work, in danger of being forgotten until Oparin summarized it, eventually led to the new field of prebiotic chemistry.

In his books, Oparin also introduced another important concept, that of *coacervates*. At first he imagined these primitive structures to be colloidal gels, having a structure similar to the protoplasm of living cells. Later, as the importance of membranes in living cells became apparent, he changed his mind. He began to envision them as tiny droplets of fatty materials that could enclose and protect collections of complex and fragile organic molecules. He suggested that coacervates might have formed naturally in the primordial soup on the early Earth. The result would have been a cloudy suspension of these fatty droplets in pond water or seawater.

These suspensions of droplets would not have been like the colloidal suspensions with which we are familiar, such as those found in homogenized milk, in which the fat has been broken up mechanically into tiny globules. They would have had much more interesting properties. Oparin supposed that coacervates were able to function as primitive cells.

The boundary of each droplet would have been made up of a layer of molecules that would have acted rather like the membrane that surrounds a living cell. Inside the coacervates and protected from the rigors of the world outside, molecules of various types could persist and interact with each other. Some of these interactions would have resulted in enhanced stability of the coacervate structure, allowing some of these little droplets to survive longer than others in the natural environment.

Some of the coacervates could even have developed the capacity to divide. Oparin assumed this ability as he expanded on his idea of chemical evolution:

> The daughter droplets resulting from mechanical division have the same physico-chemical organization as the mother droplet, since they represent portions of the same original system. They could each undergo changes which would either increase or decrease their chances in the growth competition. This would result not only in a gradual increase in the mass of organized substances on the Earth's surface, but, and this is even more important, the quality of the substance would change in a very definite direction.

The idea of a prebiotic soup, coupled with some kind of chemical selection, has dominated thinking about the origin of life on Earth ever since it was first developed over seventy years ago. It is, as we will see, a testable hypothesis, although thirty years were to pass before the first tests were carried out.

Making a Primordial Soup

The ideas of Oparin and Haldane were finally experimentally confirmed in the spring of 1953. The announcement came during a dramatic lecture given in an old and musty room with creaking floors in Kent Hall at the University of Chicago. Only the most distinguished scientists, many who either had or would eventually be awarded Nobel prizes, were usually invited to present these distinguished lectures. But this day was different—a rather nervous twenty-three-year-old second-year graduate student, Stanley Lloyd Miller (Figure 2.3), was talking. Nevertheless, the room was full because the word had spread that some new earth-shaking results were going to be presented. In the audience were some of the twentieth century's greatest scientists, who had come to the university as part of the Manhattan Project and stayed on after the war.

The topic that Stanley Miller discussed before this high-powered audience was the synthesis of important biological compounds using conditions thought to have existed on the primitive Earth. He reported that by sending repeated electric sparks through a sealed flask containing a mixture of methane, ammonia, hydrogen, and water vapor, he had made some of the amino acids found in proteins. Perhaps, he suggested, this was how organic compounds had been made on the ancient Earth before life existed.

FIGURE 2.3 Stanley Miller at a University of Chicago Chemistry Department Christmas party, 1953. (Courtesy Stanley Miller)

Although Miller was confident of his results, the rows of famous faces in his audience were, to say the least, intimidating. He was bombarded with questions. Were the analyses done correctly? Could there have been contamination?

The late Carl Sagan, then an undergraduate at the University of Chicago, was in the audience. He recollected afterward that Miller's inquisitors seemed picky and did not appreciate the significance of the experiment. Even the relevance of Miller's results to the origin of life was questioned. At one point, someone asked how Miller could really be sure this kind of process actually took place on the primitive Earth. Nobel Laureate Harold Urey, Miller's research adviser, immediately replied, "If God did not do it this way, then he missed a good bet." The seminar ended amid the laughter, and the attendees filed out with some making complimentary remarks to Miller.

The original idea for this landmark experiment had actually been suggested eighteen months earlier, in the fall of 1951. Urey (Figure 2.4) had given a lecture, in the very same room where Miller was now presenting his extraordinary results, dealing with the origin of the Solar System. He traced for his audience how the process of planetary accretion must

FIGURE 2.4   Harold Urey (1893–1981) at the University of Chicago circa 1950. He is sitting at one of the first mass spectrometers used for the determination of carbon isotopes in natural samples. From the Harold Clayton Urey Papers, Mandeville Special Collections Library, University of California, San Diego.

have given rise to certain chemical conditions on the early Earth and how these conditions were probably important to the origin of life. He summarized the notion of a primitive atmosphere in which chemical reduction, the opposite of oxidation, could have been the predominant kind of chemistry. And he explained that given recent advances in the understanding of geochemistry and cosmochemistry, it was very reasonable to suppose that the primitive Earth had a reducing atmosphere.

A reducing atmosphere, as Urey envisioned it, would have been one containing plentiful quantities of energy-rich gases such as hydrogen, ammonia, and methane. These "reduced" gases could have reacted together to produce organic compounds. Such an atmosphere would have stood in vivid contrast to the oxidizing atmosphere that our planet now possesses, with its plentiful free oxygen that destroys rather than builds up organic materials.

Lying between these extremes is another possibility, a chemically neutral atmosphere consisting primarily of inert, or nonreactive, gases. An example of a neutral atmosphere would be one made up of nitrogen and carbon dioxide. In a world with such an atmosphere, organic compounds would not be broken down, but neither apparently would they be synthesized.

Up until that time, Urey pointed out, the scientific community had conducted few experiments to try to mimic prebiotic organic synthesis. One experiment had been performed by Melvin Calvin, a biochemist who was fascinated by the process of photosynthesis and would later win the Nobel prize in 1961 for working out important parts of this fundamental biological mechanism. In 1950, Calvin and his co-workers tried to see whether photosynthesis-like reactions could take place in the absence of living organisms. They irradiated carbon dioxide and water with high-energy helium ions generated by a cyclotron at Berkeley, California, and managed to obtain formic acid and tiny amounts of formaldehyde, a first step in the formation of sugars.

Urey remarked acerbically that "if you have to go to these measures to get organic compounds, then perhaps a new idea is needed." He suggested that someone needed to try to synthesize organic compounds using reducing conditions, which so far had never been done, even though Oparin had suggested the idea some twenty-five years earlier.

Stanley Miller was just beginning graduate studies in the Department of Chemistry in the fall of 1951 when he attended Urey's lecture. Fascinated, he still remembers aspects of the lecture in great detail. But he did not immediately jump at the opportunity to do the experiment that Urey had suggested. Instead he decided to concentrate on theoretical work.

He began a project with Edward Teller, the father of the hydrogen bomb, to determine how chemical elements might have been synthesized in the early universe. After nearly a year had gone by without much progress (the problem was actually in the process of being solved in elegant detail by Geoffrey Burbidge, Margaret Burbidge, Edward Fowler, and Fred Hoyle), he began to think again about Urey's talk. He approached Urey in September 1952 about the possibility of doing a prebiotic synthesis experiment using a reducing gas mixture.

Urey was not very enthusiastic. He felt, with some justification, that graduate students should do experiments that had a reasonable chance of working, rather than taking a leap into the unknown. He suggested instead that Miller work on determining the amount of the element thallium in meteorites, a safe and pedestrian topic. But Miller was persistent. Urey finally relented and let him try some experiments, but specified that there must be signs of success within a year or the project should be abandoned.

The first challenge was to design an experiment. The mix of water and gases that Miller wanted to try was unlikely to do anything interesting if it just sat there in a flask. He needed some sort of high-energy input to encourage chemical reactions.

Miller knew that chemists had been experimenting with electric sparks in gas mixtures since the end of the nineteenth century, sometimes producing interesting syntheses, but it seemed that no one had thought about how this might relate to prebiotic syntheses and the origin of life. He realized that electrical discharges were probably common on the early Earth. The atmosphere then must have been filled with lightning flashes and with auroral displays far more spectacular and energetic than the present-day northern lights.

Calvin had focused on trying to synthesize carbohydrates, but Miller thought amino acids would be much more interesting. One could not make a living organism entirely out of carbohydrates, but amino acids could join together in chains to make proteins, and proteins were among the most important constituents of living cells.

Making a World out of Glass

What kind of apparatus would be needed for such experiments? Together, Miller and Urey sketched a diagram showing two glass flasks connected by two glass tubes. The small flask would contain water, which could be heated. The other, larger flask was designed to have two electrodes projecting from the outside into the interior, between which sparks could leap (Figure 2.5).

One of the connecting tubes was surrounded by a condenser, a coiled tube that could be cooled when water passed through it. This allowed vapors to condense out of the flask with the mix of gases, dribble down the coil, and drain back into the main flask. The second tube had a valve or stopcock, which Miller could use first to connect the apparatus to a vacuum line to remove all the air and then to introduce various gases into the system.

This design was meant to simulate the ocean–atmosphere system on the Earth. Water vapor produced by heating the main flask would be like evaporation from the oceans. The gases and water vapor in the

FIGURE 2.5   The glass apparatus used in Miller's experiment. The lower flask was designed to simulate the oceans; the upper flask, the atmosphere. Energy was supplied by sparking between the two wire electrodes. (Scripps Institution of Oceanography)

flasks would act as a tiny atmosphere, and the sparks in the flask with the electrodes would simulate lightning. The condenser would collect any newly synthesized compounds from this mini-atmosphere, allowing them to be washed into the "ocean" (the flask with the water).

It took about a week for the glassblower to build the apparatus. The stage was set for the first experiments. Before any experiments could be started, however, all the oxygen had to be removed from the system. This was important for two reasons. First, there had certainly been no free oxygen on the primitive Earth—all the oxygen that we currently breathe has been produced by the activities of green plants. Second, Miller and Urey were all too aware that sending an electric spark through a mixture of gases such as hydrogen and methane in the presence of oxygen could generate an explosion, one of the great fears of any chemist!

After pumping all the air out of the apparatus, Miller introduced a mixture of methane, ammonia, and hydrogen. He started the spark and began to heat the water flask gently. After two days, the water had turned pale yellow and a tarry residue coated the inside of the atmosphere flask around the electrodes. Anxious and unable to wait any longer, he stopped the experiment at that point and analyzed the water for amino acids.

The analytical methods that Miller had available to him were rather primitive. He separated the compounds in the water flask using paper chromatography, a method in which the bottom of a suspended sheet of paper is dipped into a solvent mixture. As the solvent creeps up the paper, it carries with it the various molecules in the sample, moving distances up the paper.

He then dried the sheet and sprayed it with a compound called ninhydrin, which reacts with different amino acids to give a variety of purple, red, and yellow tones. (Ninhydrin is also used by criminologists searching a crime scene, because fingerprints leave a smudge of amino acids.) To his great excitement, a single, faint purple spot showed up on the paper, just in the position where glycine, the simplest amino acid, should have been.

While Miller was conducting these experiments, Urey was out of town on a lecture tour. During one lecture, he mentioned that he had a student who was testing the reducing-atmosphere prebiotic-synthesis idea. Someone in the audience then asked, evidently not too politely, "And what do you expect to get?" Urey replied shortly, "Beilstein." He was referring to the one-hundred-plus-volume compendium begun by Friedrich Beilstein, *Beilsteins Handbuch der Organische Chemie*, which describes all the organic compounds that have been synthesized. And that one-word answer showed that Urey was now very optimistic about the chances for success with the experiment, a big change from only a few weeks earlier. What he did not predict was that some compounds, particularly amino acids, would be far more plentiful than others.

When Urey returned from his trip, he was naturally very pleased when informed about the glycine spot. Miller repeated the experiment, this time sparking the apparatus for a week and boiling the water rather than heating it gently. At the end of the week, the inside of the sparking flask was coated with an oily scum and the water solution was yellow-brown. Now the glycine spot on the paper chromatogram was far more intense, and spots corresponding to several other amino acids were showing up as well.

Miller estimated that he had made amounts of these amino acids that could be measured in thousandths of a gram, surprisingly abundant considering that the synthesis conditions were so nonspecific. All this had been accomplished in a little over three months, well within the time limit given by Urey.

After he showed the new, impressive results to Urey, they decided that it was time to get them published, preferably in a leading journal such as *Science*. Urey called the editor, Howard Meyerhoff, and asked that the paper be published as soon as possible—an approach open to Nobel laureates but to few others. Meyerhoff replied that this could be done in about six weeks. Miller wrote a draft of the paper. When he showed it to Urey, he was surprised by Urey's immediate and generous response. Urey said that his own name should not be on the paper, because if it were, Miller would receive little or no credit.

The paper was submitted in mid-December 1952. After waiting for the promised six weeks, Miller still had heard nothing from the editor about the status of the paper. When Urey found out about this, he was furious. He had Miller withdraw the paper and submit it elsewhere. After Meyerhoff called frantically, however, promising to get it out right away, Miller resubmitted the paper, which was published in *Science* on May 15, 1953. The delay was caused by a reviewer who had refused to believe the results and who had put the paper aside without sending any comments back to Meyerhoff.

Miller had just turned twenty-three and was about to become famous. The worldwide reaction startled him. The *New York Herald Tribune* ran a story the day the paper appeared in *Science* with the headline "Test Backs Theory Life Began as Chemical Act: Scientist, Twenty-three, Hailed." The *New York Times* published an editorial the same day titled "Life and a Glass Earth." The piece was cautious about the achievement, however: "He actually synthesized some amino acids and thus made chemical history by taking the first step that may lead a century or so hence to the creation of something chemically like beefsteak or white of egg. Miller is elated, and so is Professor Urey, his mentor."

Time, *Newsweek*, *Life*, and *Scientific American* all carried stories about the experiment. Pictures of Miller with the glass apparatus were flashed throughout the world. Nonetheless, even during this intense coverage and interest, when a Gallup poll asked adults throughout the United States, "Do you think that science will ever be able to create life?" a resounding 78 percent said no. The public's caution was justified, of course—Miller's experiment was a long way from the creation of life.

PRIMORDIAL SOUP

The attention of the public is fleeting, and things soon returned to normal. Miller started to refine the details and the analytical aspects of the experiment. The first order of business was to make sure that his compounds really were what he thought. He used melting point determinations, which at that time were considered the most conclusive way to identify organic compounds. These tests confirmed the identities of the amino acids that he had found earlier and also showed that he had synthesized an even wider variety of amino acids than he had first supposed. At the end of all this painstaking work, nine different amino acids had been positively identified, and a host of others whose identity was uncertain were also shown to be present. Some that had been identified, like glycine, alanine, and glutamic acid, were found in proteins, but others were not.

Evidently because of prodding by some doubters, Urey then became worried that bacteria might have grown in the apparatus. Miller again filled his flasks with the gases and water and then placed the entire apparatus in an autoclave, similar to those used in hospitals to sterilize surgical instruments. Sterilization is normally achieved in twenty minutes under such conditions. But because he wanted to make very sure that every organism had been killed, he left the autoclave on for eighteen hours. When the spark experiment was carried out again in this thoroughly sterilized apparatus, he found the same set of amino acids as before. The amino acids were clearly not an artifact of bacterial contamination.

There had to be a good chemical explanation for Miller's remarkable results. He was aware of the classic nineteenth-century experiment of Adolph Strecker, who had synthesized alanine from a mixture of hydrogen cyanide, acetaldehyde, and ammonia. He thought that a similar synthetic process might be taking place in his own experiment, even though he had not added any hydrogen cyanide or aldehydes to his flask.

Indeed, when he analyzed the contents of the water flask, he found that large quantities of hydrogen cyanide, aldehydes, and ketones were in fact present. They had been generated by the spark discharge, and as soon as they were flushed out and accumulated in the flask containing the water, they could react with each other. Although the final step in his experiment, the one that led to the appearance of amino acids, was indeed the same as the reactions that Strecker had carried out a century earlier, something new and exciting had happened in Miller's flasks. By providing an energy source, he had manufactured the highly reactive building blocks that made the subsequent Strecker-like syntheses of more complex molecules possible.

In 1972, Miller and his collaborators repeated the experiment and used better analytical methods to detect the amino acids being formed. This time they found thirty-three different amino acids, including over half of the twenty commonly found in proteins. As they expected, amino acids that had the most carbons in their side chains were the rarest.

Amino acids were not the only compounds being produced in the discharge apparatus. Miller found another class of closely related compounds, called hydroxy acids. The simplest of these was glycolic acid, which is used in skin creams and is the hydroxy acid cousin of glycine. The hydroxy acid relative of alanine, lactic acid, was also found, as were the hydroxy acids corresponding to many amino acids that had been produced in the experiment.

But the main material that accumulated during all the experiments was the tarry, gooey stuff that coated the insides of the sparking flasks. A soluble component of this goo, dissolved in the water flask, turned it the color of strong coffee. Most of this gooey material was insoluble in water. When treated chemically, it also released amino acids. It turned out to be an incredibly complex and intractable network of organic molecules that had been linked together in innumerable different ways. As we will see, this tarry material also had its equivalents in the actual prebiotic world.

It was not long before other laboratories repeated Miller's experiments, using a variety of conditions and energy sources. Their results demonstrated how important it was to have energy-rich reducing gases in the atmosphere flask of the experiment; if methane was replaced with carbon dioxide and ammonia with nitrogen, no amino acids were produced. And they showed that other energy sources such as ultraviolet light also gave the same results, though the yields of amino acids were far lower than those obtained with a spark discharge. In the absence of a concentrated energy source, however, even if the atmosphere were reducing and the flask was boiled, nothing would happen.

The Concentration Problem

Miller's experiment was a good beginning, but it was only a beginning. Although amino acids form the building blocks of proteins, and hydroxy acids such as lactic acid are important in metabolism, to make life, one would need a much larger array of compounds than this. Some of the most important of these are the components of DNA and RNA, which make up the genetic material of present-day cells. These nucleic acids contain the bases adenine, guanine, cytosine, thymine, and uracil, and the sugars ribose and deoxyribose. If Miller's early experiments had produced some of these compounds, the level of excitement generated in the scientific community would have been even higher.

Alas, in spite of careful analysis, Miller did not find any evidence for the presence of these other key compounds. But his finding of hydrogen cyanide in the experimental mixture set the stage for further experiments. One of the most important of these, carried out in the early 1960s by John Oró at the University of Houston, showed that when a highly concentrated hydrogen cyanide solution is heated gently, a surprisingly large quantity of adenine is produced.

This was an immensely important discovery. Not only is adenine a nucleobase component of DNA and RNA, it is also an important part of the most common high-energy compound found in all living cells, adenosine triphosphate (ATP) (Figure 2.6). Virtually nothing can take place in a living cell without the help of ATP, which carries high-energy phosphates capable of donating energy to many different chemical reactions in the cell.

The pathways in which ATP plays a role include the construction routes by which living organisms make amino acids. The chemistry of living cells is far more sophisticated than the chemistry of the primordial soup, because cells are able to use a wide variety of enzymes to build amino acids from simpler compounds. ATP plays an important role in all these syntheses by donating energy at many points along the pathways.

As if this were not enough, ATP also plays an essential role in putting together the huge protein and nucleic acid molecules that are the major components of the cell. It does this by aiding in the linking together of amino acids into protein chains and by donating the energy that permits base-containing subunits (called nucleotides) to be joined together into the long chains of DNA and RNA. Adenine is also found in various other molecules that donate energy to reactions and that aid the function of many different enzymes.

It was fascinating that such an important compound as adenine was so easily synthesized under prebiotic conditions. But why had Miller failed to find it in his original experiment? After all, he had shown that hydrogen cyanide was present.

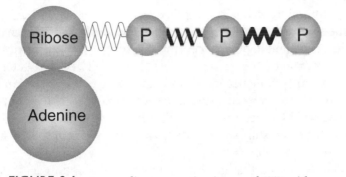

FIGURE 2.6 A very diagrammatic picture of ATP, with its three phosphates attached to the ribose. The springlike connections illustrate in a pictorial way the energy stored in the molecule.

Adenine is simply five hydrogen cyanide molecules linked together. Oró found that hydrogen cyanide polymerizes with itself in a series of reactions involving progressively more and more elaborate intermediates. The remarkable feature of this reaction is that it can take place spontaneously under physical conditions that are not at all extreme. Gentle heating is all that is required to make the reaction go rapidly.

The reason that Miller did not produce adenine in his spark discharge experiment was that the hydrogen cyanide concentration in his water solution, while substantial, was too low for detectable amounts of adenine to be produced. In small quantities, hydrogen cyanide tends to be converted into other compounds. As a result, some chemists and geologists dismissed the importance of the Oró reaction, claiming that there would never have been enough hydrogen cyanide in the primordial soup to produce meaningful amounts of adenine.

Nonetheless, the robustness and spontaneity of the Oró reaction is remarkable. Further, some conditions that might have been present on the early Earth could have led to the formation of adenine. The gentle heating that Oró had applied to his hydrogen cyanide solution simply accelerated a process that is quite able to happen spontaneously even at lower temperatures. In the late 1990s, Miller and his collaborators investigated dilute hydrogen cyanide solutions that had been sitting in their laboratory freezers for years and found that adenine had been produced in significant yields.

This freezer reaction takes place because, when dilute hydrogen cyanide solutions are frozen, most of the water freezes into ice crystals. This leaves behind highly concentrated hydrogen cyanide brines that do not freeze except at temperatures less than −20°C. As the water freezes the hydrogen cyanide becomes more and more concentrated in the unfrozen brine. You can see a similar effect when you open a container of fruit juice that has been taken straight from the freezer. You will find it filled with ice crystals surrounded by a highly concentrated and unfrozen sugar solution.

Miller also found that glycine and a few other amino acids could be produced from hydrogen cyanide solutions under these freezing conditions. And recently, Miller, Oró, and Matthew Levy have shown that small amounts of another important constituent of RNA and DNA, guanine, can be formed at a wide variety of temperatures if the experiment starts with concentrated hydrogen cyanide.

If there were times when the primordial soup froze, then high concentrations of hydrogen cyanide and other compounds in the resulting slush could have played an important role in the synthesis of several key organic molecules. If the primordial soup sometimes turned into a kind of primitive sherbet, this might have increased the richness of its ingredients!

False Starts in the Move Toward Greater Complexity

In 1958, Sidney Fox, a protein chemist then at Florida State University, announced a dramatic discovery. He thought that instead of lightning, heat energy may have been the major driving force behind many of the reactions needed to make organic compounds on the early Earth. He baked a mixture of amino acids on a hot surface at 170°C and found that he could make them join together, or polymerize. Then, when the mix of the resulting proteinlike molecules was again dissolved in water, strange structures appeared that seemed to have some of the properties of living cells.

Fox claimed that he was well on his way to creating life in the laboratory. When the German scientist Wolfram Thiemann of the University of Bremen later asked Fox, during a casual conversation in a bar, whether he had in fact created life, Fox gave a rather Jesuitical reply: "Of course I did—if you define life the way I do!"

Fox called his proteinlike substances proteinoids and claimed that their synthesis was analogous to the reaction by which a living cell joins amino acids together to make protein chains. In a living cell, amino acids are linked together by the formation of peptide bonds, in which the basic, or amino, part of one amino acid is attached to the acidic, or carboxyl, part of another. During the course of this linkage, a molecule of water is released. Chemists refer to this type of reaction as dehydration.

In a water solution, dehydration reactions are very unfavorable; when there is a huge excess of water molecules in the vicinity, it is difficult to coax amino acids to join together and release yet another molecule of water. Thus, in an aqueous solution of amino acids, peptide bond formation does not readily take place. The enzymes in a living cell can change these odds by creating water-free zones inside membranes and by adding lots of chemical energy to the amino acids in the form of the ubiquitous ATP. Baking a dry mixture of amino acids changes the odds too, though in a much less precise way.

In a living cell, amino acids join together in long chains to form proteins. In Fox's reaction, however, the amino acids could also link together in many other ways. Some amino acids have extra amino and carboxyl groups, and when these participate in the joining process, the result is branched rather than linear amino acid chains. Some amino acid molecules are altered by the intense heat of the baking process, and their decomposition products can also participate in the reactions. The result is an unpredictable mix of products.

Fox's experiment was an intriguing one, but unfortunately he stepped beyond the boundaries of his experimental data. He claimed that as his amino acids were heated and joined together, they tended to order themselves into reproducible sequences rather than simply linking up at random. He characterized this process as the very beginnings of life, constituting a "biomacromolecular big bang."

Only a few years before Fox's experiments, Fred Sanger and his co-workers at Cambridge University had painstakingly sequenced a real protein for the first time. They found that the one-hundred-plus amino acids of the protein insulin were joined together in a very specific order. It soon became apparent that all of the billions of different types of proteins in the natural world are characterized by a particular ordering of the amino acids that make up their chains. They owe their specificity and activity to this precise ordering. If Fox was right, then at least some of this ordering could have taken place even before life appeared, perhaps when amino acid solutions dried into scum on the warm rocks that surrounded the primitive ocean.

PRIMORDIAL SOUP

Fox and his co-workers then tried to imitate what might have happened when the tide washed over the proteinoids synthesized on hot rocks on the primitive Earth. They found that when proteinoids were suspended in water, they came together to form tiny, hollow spheres. Perhaps, they thought, these microspheres might be similar to the coacervates that Oparin had postulated decades earlier. Proteinoid microspheres could have evolved by chemical selection into the first self-reproducing cells.

Subsequently, Fox made a career out of investigating the properties of his proteinoid microspheres. He eventually advanced and championed the claim that these were indeed primitive cells. According to Fox, the microspheres had a wide range of catalytic activities, were made up of double layers of the kind that surround living cells, and underwent budding or fission in a manner that looked like reproduction. They also tended to assemble in long strands that resembled some of the earliest fossils that had been found in ancient rocks.

Fox was an excellent self-promoter. His claims that proteinoid microspheres were lifelike eventually appeared in several biology textbooks, something that he constantly mentioned in his talks and papers. He convinced the young government agency NASA (National Aeronautics and Space Administration) that his research was fundamental in determining how life began on Earth. He was rewarded with generous NASA funding that allowed him to set up the Institute of Molecular Evolution at the University of Miami.

In 1963, Fox organized the second international conference on the origin of life in Wakulla Springs, Florida. Oparin and Haldane came to the conference and met for the first and only time. Many other scientists, such as Oró and Sagan, who were or would become leaders in the origin-of-life field, also attended.

Alas, the conference was heavily stacked to promote Fox's proteinoid theory. Stanley Miller was conspicuously absent. Miller considered Fox's claims to be exaggerated, whereas Fox felt that Miller's experiment was only a minor step toward solving the origin-of-life problem.

Fox's work turned out to be too good to be true. Soon holes began to appear in his claims. In particular, a number of scientists examined his assertion that amino acids tended to join together nonrandomly when they were dried, and found that it was not correct. Proteinoids really were the result of essentially random linkages among amino acids. And it was difficult to see how such distorted, nonlinear, and indeed nonbiological molecules could have played much of a role in the origin of life.

Although Fox continued to champion his ideas through numerous papers, popular articles, books, and talks until he died in the summer of 1998, his work is now almost universally dismissed. Over time, he became more and more a maverick in the field. Sadly, at the Eleventh International Conference on the Origin of Life, held in Orleans, France, in 1996, he was reduced to having placards of proteinoid microspheres paraded around in the manner of a cartoon sandwich man predicting the Second Coming.

The scientific community may have reacted too harshly. Fox did show that amino acids could join together into larger molecules under conditions that might have taken place on the primitive Earth. And he did show that the resulting compounds formed interesting and complex structures, although they certainly did not have the lifelike properties that he attributed to them. The possibility that simple peptides that formed in this fashion, perhaps so short that they did not branch, could have played an early role in the origin of life cannot be dismissed completely.

But because of his unsubstantiated claims, most scientists have avoided further experimentation with proteinoids. They have instead turned to other and more reproducible methods for linking amino acids together under prebiotic conditions. As we will see, many of these are turning out to be very exciting—and these new results are being presented in a far more cautious fashion.

Another scientist who ran into difficulties because he made sweeping claims was Cyril Ponnamperuma, from Ceylon (now Sri Lanka). Ponnamperuma began his work on the origin of life in the laboratory of Melvin Calvin at Berkeley, then moved to the NASA Ames Research Center. There he eventually became head of the Life Synthesis Branch, later the Chemical Evolution Branch, which was involved in the analyses of rocks and soils returned from the Moon during the Apollo missions. He also continued to do prebiotic synthesis experiments, using the experimental design originally used in Calvin's 1950 experiment.

In 1963, Ponnamperuma reported that adenine was produced by the bombardment of reduced gases with electrons, and in collaboration with Carl Sagan, he announced that he had synthesized ATP under simulated prebiotic conditions. He claimed in 1968 that compounds called porphyrins had been produced in the electron bombardment experiments as well. This would be a particularly important finding because chlorophyll, the photosynthetic pigment of plants, is a porphyrin. Unfortunately, all these results have proven difficult to confirm, and most researchers now view them with skepticism.

Many early claims of movement toward greater chemical complexity from the molecules found in Miller-type experiments have tended to fall apart on subsequent examination. Why has it been so difficult to produce complex molecules in prebiotic experiments? One problem seems to be simply a matter of probabilities. Another is the difficulty of detecting compounds at very low concentrations.

Fox's proteinoids are a good example of this. An experiment in which amino acid solutions are dried and then dissolved in water again will produce a great variety of proteinoids. The problem lies in that very variety—there are so many different ways to join a mixture of amino acids together into chains!

Suppose we start with a mix of ten different amino acids. Then the number of possible different peptide molecules that we could make, each consisting of a chain with a length of ten amino acids, is ten raised to the tenth power (10^{10}), or 10 billion. And this assumes that the amino acids are joined together in simple chains and takes no account of all the possibilities for branching.

Indeed, so great is the number of possibilities, that most of the proteinoid molecules that Fox made in his baking experiments were probably unique. If a proteinoid with interesting properties did happen to be formed by chance, it would be lost in the huge and diverse crowd of other proteinoid molecules.

The diversity problem is confounded by the fact that molecules are very small, and there are a great many of them in even a small sample. For example, suppose a tiny amount of a key molecule required for the origin of life was actually produced in the prebiotic experiments that Miller and others carried out. Could it have been detected? Using sophisticated methods, chemists can often detect as little as a trillionth of a gram of a compound in a milliliter of a solution. But even this minute amount is made up of billions of molecules. Thus, even if the solution were to contain a few million molecules of this key component, it might easily be missed. The problem would be far worse if there were only a handful of these key molecules—there is presently no way that such tiny quantities can be detected.

The chemical processes taking place in the prebiological world must have sometimes led to organic molecules of substantial complexity that had important catalytic or information-carrying capacity. But in the absence of some sorting-out principle, these molecules would have been simply trace components in an immense zoo of other molecules, resembling the *Beilstein*-like mixture that Urey had predicted to be the outcome of Miller's experiment. Some mechanism was required for sorting out this mixture and enriching the mix for certain kinds of molecules with specific properties important in the origin of life. That mechanism, as we will see, must have been a Darwinian one.

Working Toward the Golden Spike

The building of the first transcontinental railroad in the United States, the Union and Central Pacific, was one of the great industrial achievements of the nineteenth century. Workers starting out from Omaha, Nebraska, in the east and Sacramento, California, in the west met in triumph at Promontory Summit, northwest of Ogden, Utah, on May 10, 1869. There they drove in a golden spike to mark the railroad's completion.

Like those gangs of railroad workers, science is working in two directions toward the origin of life. But rather than starting from the east and the west, they are working from the top downward and the bottom upward. The point at which these two groups of scientists meet in the middle will be at least as epoch-making an event as that meeting in Utah. Like the completion of the railroad, that event will change our lives forever.

A successful completion will involve two approaches. The bottom-up approach to the origin of life is focused on the creation of some kind of self-replicating entity in the laboratory, starting with simple substances under some approximation of prebiotic conditions and ending up with a structure that has at least some properties of life. The first challenge will be to find new and sensitive ways of detecting these entities once they have appeared in the course of an experiment, for they are likely to be few and feeble. The next challenge will be to persuade other scientists that this goal has really been accomplished. Because such "life" is likely to be chemically different from today's living organisms, the interpretation of these experiments will be hotly contested.

The second approach, from the top down, will succeed when it is possible to dissect modern life to its essentials, demonstrating the steps by which the elaborate machinery of the living cells of today first appeared. Huge challenges lie in the way of accomplishing this second goal as well. The genetic code is complex, and it is not obvious how it could have started out as some simpler set of instructions. The machinery for putting proteins together from amino acids is very elaborate, carried out on structures called ribosomes, which are huge clusters of cooperating molecules. Even the machinery for making the essential high-energy compound ATP is astoundingly complex—the major enzyme involved is a Robocop-like structure made up of seven large subunits, one of which rotates like the giant piston of a rotary-engine car. How did such complicated structures arise from simple beginnings? Luckily, we will see that there are clues to their history that are buried in all these complicated systems.

We predict that sometime early in this century, the top-down and bottom-up lines of investigation will indeed manage to meet. Whether there will be a golden spike ceremony to mark the occasion, with the successful teams shaking hands and grinning at the camera, remains to be seen.

We will spend the next few chapters exploring the bottom-up approach, which is continuing to yield remarkable findings. Later we will explore the top-down approach, which too is producing striking insights. Although the gap between the two approaches still yawns, we can begin to see how it might be bridged.

Since the time of the Miller-Urey experiments, chemists have succeeded in producing many different organic molecules under plausible prebiotic conditions. In many cases, however, this has become a goal in itself, with each success trumpeted as if it were a further step toward understanding the origin of life. But these experiments, fascinating as they are, do not tell us how life originated. What they do tell us is that there is something inherent in the way that chemistry works, something that can readily lead to molecules that could form a basis for life. The alphabet, it seems, was present in the primordial soup. The trick is to make a great novel from it.

But even getting to the alphabet might have been more problematic than scientists had supposed during the first flush of enthusiasm following Miller's experiment. Problems soon arose with the Miller-Urey view of the world. These problems had nothing to do with the chemistry, but rather with the basic assumptions that underlay their experiment. To understand these problems, we must reconstruct as well as we can how the Earth formed and what the conditions on our primordial planet might really have been like.

3

The Earth's Apocalyptic Beginnings

[H]e stood before a phenomenon which revealed to him the proximate origin of all the stellar heavens.

John Pringle Nichol (1836), discussing the ideas of William Herschel

In 1993, J. William Schopf, at the University of California at Los Angeles, demonstrated unequivocally that tiny chains of single-celled organisms lived on the Earth 3.5 billion years ago. He found these microscopic fossils, visible only under the microscope, inside ancient rocks from western Australia (Figure 3.1). They could be seen only when the rocks were painstakingly polished down to layers so thin that light could be shone through them. After this polishing, the wafer-thin films of rock glowed under the microscope like the thin sheets of mica in the windows of an old pot-bellied stove.

Along with the rocks in which they were now firmly embedded, the cells had undergone many changes in the process of fossilization. They might easily have been missed: Two students had failed to see anything in this material before Schopf discovered these distorted, fragmented, but still recognizable ancient chains of cells.

These were not the first such old single-celled fossils to be found. Other scientists had discovered better-preserved fossils of the same type concealed in rocks ranging from 2 billion to less than 1 billion years old. But Schopf's tiny structures were by far the oldest signs of life. They long predated the first multicellular organisms, which put in their first appearance in the fossil record a mere 600 million years ago.

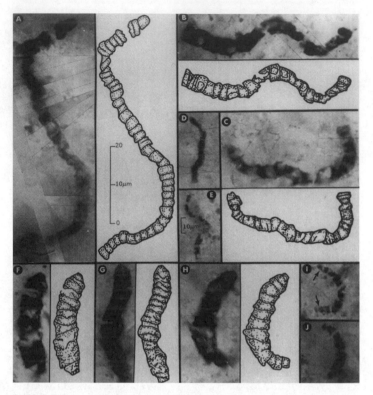

FIGURE 3.1   Some of the fossils found by Bill Schopf in
Australian rocks 3.5 billion years old. Even by this early date,
cellular structures can be seen clearly. (Courtesy J. William Schopf)

Schopf's fossils have preserved enough detail to show that by as early as
3.5 billion years ago, life had advanced well beyond the very primitive
stage. The organisms living then looked very much like the common
photosynthetic bacteria that live in the oceans today. And the modern appearance of the ancient fossils meant that these tiny fossils must already
have had a long evolutionary history. They were the oldest fossils yet
found, but they were apparently far from being the oldest living entities.

Gustaf Arrhenius (the grandson of Svante Arrhenius) and co-workers
at the Scripps Institution of Oceanography, as well as others have claimed
that rocks from a remote mountainous region at the edge of the massive
Greenland ice sheet contain evidence that life existed 3.8 billion years ago.
The evidence is not based on actual fossils, but rather relies on indirect
carbon isotope studies. As a result, the findings are highly controversial.

Nevertheless, if this claim is eventually verified, it implies that the very first life originated even earlier than 3.8 billion years ago. But how? Did Miller-Urey-type reactions generate a rich prebiotic soup from which life emerged, as Oparin and Haldane had postulated? What if the conditions on the early Earth were unsuitable for "home-grown" prebiotic synthesis to take place readily? Would a prebiotic soup even have been present on the early Earth?

This last vexing question was raised by the distinguished nuclear physicist Philip Abelson. Abelson has taken many iconoclastic positions, which he was able to publicize during his twenty-two-year editorship of the journal *Science*. During the Apollo project, he testified vigorously to Congress that it made no sense to send men to the Moon—robotic probes could do the job far more cheaply and safely. His comments were all bitingly true, but distressingly unpatriotic during a time when the United States was in a heated contest with the Soviet Union to be the first to land astronauts on the Moon.

In 1966, when Abelson was director of the Geophysical Laboratory at the Carnegie Institution of Washington, he became one of the first to question the assumptions behind what he called the "model experiments" of Miller and others. He pointed out that the Earth's early atmosphere was probably never reducing like the atmosphere used in the Miller-Urey experiment. In particular, there could not have been much hydrogen around, because it was so light that it would have been lost quickly to space. Methane and ammonia would have broken down rapidly because of blasts of ultraviolet light from the Sun.

He contended that no matter how reducing the atmosphere might have been at the outset, it would soon have consisted primarily of carbon dioxide and nitrogen, along with some carbon monoxide and possibly traces of hydrogen. Moreover, he showed that the Miller-Urey experiment would not have worked very well in such an atmosphere. He carried out a Miller-Urey-type experiment using a mixture of carbon monoxide and nitrogen instead of their reducing gas mix. Some hydrogen cyanide and the simple amino acid glycine were produced, but apparently not much else. Abelson's experiments began a fierce debate, still going on today, about the composition of the Earth's early atmosphere.

There are several ways to escape from the difficulties that Abelson raised. One is to assume that in spite of the various concerns he and others had raised, the early atmosphere somehow managed to stay reducing. For instance, reducing gases might have been supplied in quantity if intense volcanic activity had been widespread. In addition, much of the soup might not even have formed on the Earth. Perhaps its ingredients had been created elsewhere in the Solar System and had showered down on the early Earth as a constant rain of carbonaceous meteorites.

We can explore these possibilities by journeying back in time to watch the Earth and the other planets of the Solar System as they condensed out of a great primordial gas cloud. As we do, we will understand more vividly the perils faced by our first primitive ancestors, and the awesome and alien physical environment in which life first appeared.

Genesis of Our Solar System

Imagine that the time is 4.5 billion years ago and you are drifting in space, observing our Solar System in its infancy. You are on the periphery of a great disk of gas, dust, pebbles, and rocks that encircles the young Sun. The Sun is shining nearby, but it is much less bright than today's Sun, and on top of this its rays are masked by the material in the disk. Instead of the familiar planets dotted about a black sky, there is a golden, glowing haze everywhere.

The material making up the disk around our infant Solar System consists of the remnants of an earlier and much more extensive gas cloud, in the middle of which the Sun had coalesced, contracted, and finally reached a large enough mass and density to blaze up with nuclear fire. The original gas cloud had contained a plentiful inventory of the various chemical elements. The lighter elements had been synthesized inside other stars as they burned hydrogen and helium to produce energy, and the heavier elements had been produced during the huge cosmic explosions known as supernovas, of giant stars. The elemental waste products of these stellar processes were strewn throughout the universe, becoming incorporated into accumulating gaseous clouds such as the one that you now see coalescing into our Solar System.

In the heart of the disk, close to our infant Sun, it is hot enough to melt rocks. But in the outer fringes of the disk, so little of the Sun's radiation penetrates through the haze that temperatures are just a few degrees above absolute zero, the temperature at which atoms and molecules stop moving. Because of this very large temperature gradient, gases like hydrogen, along with a variety of simple compounds made up of hydrogen and other light elements, have been largely driven out of the center of the disk and into the colder, outer regions. As a result of this process, the inner disk consists mainly of the heavier elements that make up rocks, and the lighter materials far from the Sun have congealed to become chunks of "dirty" ice made up of frozen water, ammonia, methane, and other compounds.

Most of these ice balls now move in stately, circular orbits, though a few are perturbed by interactions with their fellow lumps of ice. Some are catapulted to the utterly frozen outer reaches of the Solar System, and others begin to fall toward the distant Sun.

Throughout the disk, larger objects begin to attract smaller ones, and gradually the components of the disk begin to coalesce. As you watch the Solar System mature, larger chunks of material begin to collect together into little protoplanets, or *planetesimals*. Soon, thousands of these planetesimals are moving around the Sun, each in its own orbit. Most of these orbits are only approximately circular, and as the planetesimals interact with each other, their orbits are constantly changing. In the resulting dodgem ride, collisions are numerous (Plate 2).

Some of these collisions begin to form planetary embryos, and as the embryos grow in size, their gravitational pull increases. More and more planetesimals are pulled toward these growing objects, which soon begin to move like vacuum cleaners through the dust cloud, leaving tracks of empty space behind them. As the planetary embryos grow and their gravitational fields increase in size and intensity, the fragments drawn toward them come from further and further away and begin to hit with higher velocities. These progressively more violent collisions are enough to melt the surfaces of the larger objects.

Some of the largest among these objects are destined to become the present-day planets. The planets that begin to form in the inner Solar System are rocky, but very different planets start to take shape in the regions far away from the young Sun. These farther planets are gas giants, made up of the lighter elements and compounds, and they accumulate thick atmospheres.

If you could fast-forward through this process, you would see that all this planet-building happens with surprising speed. The time for the accretion of the rocky inner planets is a mere 10 to 20 million years, whereas it takes a bit longer for the massive gas giant planets like Jupiter to form.

Recent computer simulations of this accretion process show that the inner regions around the Sun were initially quite densely populated with rocky planets. There were far more planets than Mercury, Venus, Earth, and Mars, the four that are found near the Sun today. Some

THE EARTH'S APOCALYPTIC BEGINNINGS

revolved in orbits that took them far above or below the plane of the coalescing Solar System. Over time, some of these planets were pulled into the Sun and destroyed. Others passed near other planets and were flung, like rocks from a slingshot, out of the Solar System altogether into the emptiness of interstellar space. And some smashed into each other just as the planetesimals had done earlier, but with far more dramatic consequences.

Is this how the Solar System really formed? Evidence is growing all the time that the events we have just pictured did take place—indeed, we can now use the Hubble Space Telescope and the huge new ground-based facilities such as the Keck telescope on Hawaii's Mauna Kea to watch the same thing happening around other nearby stars. The implications for the origin of life, as we will see, are profound.

The idea that the Solar System was formed from a cloud of dust and gas was first suggested and refined some two hundred years ago by the German philosopher Immanuel Kant, the French mathematician Pierre Simon Laplace, and the English astronomer William Herschel. Their "nebular hypothesis" turned out to be an inspired guess, for not only has it been one of the longest-lived scientific ideas, but new evidence continues to refine and reinforce it.

The first apparent evidence for the hypothesis, however, turned out to be a false lead. By the end of the eighteenth century several *nebulae,* or clouds of glowing gas, had actually been observed by Herschel and other astronomers. The nebulae were assumed to be solar systems in the process of formation elsewhere in the Milky Way.

It now turns out that real planetary nebulae are too small and too far away to have been seen by the telescopes of the time. The glowing dust clouds that Herschel and the others observed are far larger than any solar system. Many of these nebulae are the remnants of nova or supernova explosions, and others are vast, glowing clouds that have spread over many cubic light-years. In some cases, like the Orion nebula, these regions act as nurseries for new stars.

Even though it eventually became apparent that these nebulae were not solar systems in the process of formation, the nebular hypothesis continued to dominate thinking about the origin of stars and planetary systems. Further, the success of the nebular hypothesis has immense consequences for theories about the origin of life. The hypothesis implies that the formation of solar systems should often accompany the formation of stars and that the universe should be filled with planets on which life could arise.

FIGURE 3.2   In a star-forming region in the constellation Taurus, the bright dot in the bottom left-hand corner may be a Jupiter-like planet being ejected from a newly forming pair of binary stars. (Courtesy Susan Terebey and NASA)

By now, it is clear that planetary nebulae are everywhere throughout the universe. Remarkably, one star system in the process of planet formation may have been discovered. What may be a planet two to three times the mass of Jupiter has been caught in the act of being flung from a young binary star system in the constellation Taurus (Figure 3.2). Using the Near Infrared Camera and Multi-Object Spectrometer

THE EARTH'S APOCALYPTIC BEGINNINGS

(NICMOS) that is part of the Hubble telescope, Astronomer Susan Terebey of the Extrasolar Research Corporation in Pasadena, California, and her colleagues discovered this little spot of light with its trailing wisp of gas that leads back to its parent stars. They are not yet certain that it is actually a planet and not a brown dwarf, an object having a mass between that of a large Jupiter-like planet and a sun-like star. There is even a possibility that it is an artifact—nothing more than a background star. Additional new measurements should settle the matter soon, but whatever the outcome, it is likely that planets are indeed ejected from newly forming solar systems.

Other solar systems in even earlier stages of formation have been glimpsed by Glenn Schneider and his co-workers at the University of Arizona, also using the NICMOS. When the blinding glare of the star HR4796A was blocked out, a ring made up of immense numbers of tiny particles a thousandth of a millimeter in diameter could clearly be seen orbiting around it (Plate 3). This ring of dust is the raw material that will eventually accrete into the planetesimals that in turn coalesce into planets. HR4796A is estimated to be a mere 10 million years old, and its solar system is forming swiftly, exactly the kind of process predicted by the nebular hypothesis.

And there is more indirect but nonetheless convincing evidence that our galaxy is crowded with solar systems. Studies of gravitational perturbations of nearby stars suggest that at least one in every ten Sun-like stars have Jupiter-sized companions orbiting around them. We will explore the implications of this new evidence later.

So many extrasolar planets and collapsing dust clouds have now been found that we are continually revising upward our estimates of how many planets—and how many Earth-like planets—might exist in the Milky Way and in the universe at large. There appear to be many potential worlds in the universe where life could have arisen.

But with respect to the origin of life on Earth, we still need to know some essential things. What were the juvenile Earth and Solar System really like? What does the early geological record tell us about this ancient period of Earth history? One of the most dramatic events, only recently understood, had a huge effect on our young planet—and may have aided the formation of life.

An Infant Earth Gives Birth to a Moon!

In 1658, James Ussher, Archbishop of Armagh and Primate of Ireland, estimated that the Earth was created in the early evening of October 22,

4004 B.C. He based his calculations rather loosely on the family trees found in the Old Testament, and anchored them in historical events that seemed to have corresponding accounts in both the Bible and ancient written histories dating from Greek and Roman times.

This was not the first attempt to develop an estimate for the age of the Earth based on religious documents. The first calculation based on biblical chronology was made in A.D. 169 by Theophilus of Antioch, who derived a time for creation of 5529 B.C. It is not surprising that other religions have come up with very different estimates. In 120–130 B.C., Hindu priests, with commendable precision, estimated the Earth to have been 1,972,949,091 years old!

As we saw earlier, Lord Kelvin used the physics of his day to extend Ussher's estimate dramatically, although he fell far short of the estimate of those Hindu priests, or even of the hundreds of millions of years that Darwin's theory of evolution required. But even his calculation that the Earth was 24 million years old was a substantial underestimate of the Earth's true age.

A gradual approach to the current, and probably final, estimate was made possible by the discovery that radioactive decay could transmute an atom of an unstable parent element into one of a stable or unstable daughter element. At the turn of the century, Ernest Rutherford and his colleagues, working at McGill University in Canada, found that helium was produced from the radioactive decay of radium. The radium itself was an unstable daughter produced from uranium decay. The scientists realized that because the parent radioactive elements produced helium at a fixed rate, the ratio of helium to its parent elements could be used to date the time of formation of the rocks themselves. And the ability to date the Earth's oldest rocks would set a minimum age for the Earth.

The first calculations yielded ages of around 500 million years. Rutherford noted, however, that these were minimum estimates because some of the helium, being a light, gaseous element, could have diffused out of the rock and been lost.

At the same time, the American chemist Bertram Boltwood noted that all of the rocks that contained uranium also contained lead in addition to helium. He concluded that like helium, lead was a product of the radioactive decay of uranium. Moreover, the amounts of lead and helium were greatest in the oldest geological deposits. Using the best estimates he could make of the rate of uranium decay to lead, and the measured ratio of lead to uranium, he calculated in 1907 that some

THE EARTH'S APOCALYPTIC BEGINNINGS

rocks had ages greater than 2 billion years. Furthermore, since these were the ages of rocks that had solidified after the molten Earth had formed, the actual age of the Earth had to be substantially older.

These remarkable calculations increased the age of the Earth to a vast period of time that would have been unimaginable to Kelvin, though it would not have surprised those Hindu priests. But even these astounding estimates soon began to undergo revision, for not all the stable daughters in the radioactive decay of uranium and other radioactive elements were known at the time. It took many careful measurements for the decay rates, which were very slow, to be determined accurately.

In 1929, in his last contribution to the topic of the age of the Earth, Rutherford used the estimated decay rates of two uranium isotopes (uranium 235 and uranium 238) to calculate an age for the oldest rocks of at least 3.4 billion years. He also estimated that the Sun had to be at least 4 billion years old. He died in 1937, not knowing that even these immense ages were still underestimates.

Rutherford was not particularly interested in the larger implications of his measurements. Late in his life, as director of the prestigious Cavendish Laboratory at Cambridge University, he once snarled at the young physicist and mathematician J. D. Bernal, "Don't let me catch anyone talking about the universe in my laboratory!" But there were larger implications nonetheless, particularly concerning the age of the Solar System and the origin of life.

In the late 1930s a new instrument, called a mass spectrometer, was developed. It could determine the precise abundances of isotopes—literally by counting them atom by atom. Individual atoms were torn free from rock samples and fired through a magnetic field, in such a way that their numbers and masses could be determined directly. At the same time, the half-lives of the uranium isotopes, the time that it takes for half the original radioactive atoms to decay away, were established with much greater precision.

These advances permitted far more accurate estimates of the age of the Earth. As a result, it came to be generally accepted during the 1940s that the most ancient Earth rocks were about 3.5 billion years old. This venerable age for the Earth even received the blessing of the Vatican in a 1951 address by Pope Pius XII to the Pontifical Academy of Sciences.

However, scientists realized that most Earth rocks probably did not go back to the beginning of the planet's history, so that this was still a minimum estimate. But perhaps meteorites, if they were leftover bits of the original material that had accreted around the Sun, might provide some of the information needed to date the real beginning of our planet. In 1953, using the uranium/lead dating method and combining data obtained from meteorites and the Earth itself, Clair Patterson at the California Institute of Technology and Friedrich Houthermans at the University of Bern independently announced that the age of the Earth, and therefore the age of the Solar System itself, was about 4.5 billion years. By 1956, Patterson had refined his estimates to yield an age for the Earth of 4.550 ±0.070 billion years. This value remains essentially unchanged today.

As various meteorites were examined and dated, the time during which other parts of the Solar System had formed could also be estimated. G. Brent Dalrymple, in his book *The Age of the Earth,* lists the ages of twenty-six different meteorites as 4.550 ±0.025 billion years, a remarkably narrow range. These numbers tell us that virtually all meteorites, and by inference the Earth as well, had formed during a very short span of time. This is the primary evidence that the accretion of planetesimals and planets in the young Solar System must have been very swift.

A few younger meteorites have been found, but these are thought to be pieces of Mars and the Moon that were hurled into space by more recent impacts. The majority of meteorites come from the very beginning of the Solar System.

What about rocks on the Earth that date to its time of formation? Ever since Rutherford, there have been extensive searches for rocks that can be dated to the first half billion years of the Earth's history. Unfortunately, none have been found.

Tiny zircon grains 4.0 to 4.3 billion years old have been picked out of younger rocks in Australia and Canada. Aside from these ancient, sparkling fragments, which tell us nothing about what the Earth was like when they formed, the oldest rocks found are those from Greenland's Isua formation, and are 3.6 to 3.9 billion years old. This means that the first 700 to 800 million years—almost a fifth—of the Earth's history left no trace in the geological record. The oldest rocks have been obliterated by the constant recycling of the Earth's crust by the process of plate tectonics.

But, what about rocks on the Moon? When the first lunar samples were brought back by the Apollo astronauts, it was expected that they would be either about the same age as the Earth or a bit younger, depending on the mechanism by which the Moon formed.

Fragments of lunar rock were carefully collected during the Apollo missions to the Moon's ancient highlands, where it was thought that some old rocks would have escaped being smashed up completely by later meteorite impacts. The oldest highland rocks gave ages of between 4.50 and 4.52 billion years, some 30 to 50 million years younger than the ages of meteorites. This difference, although it seems small, is important, because it gives an essential clue to how the Moon formed.

The most likely scenario for the Moon's formation, first suggested by William Hartmann in 1975, is a truly spectacular one. He proposed that a violent collision took place between the infant and still moonless Earth and a Mars-sized planet, one of the many other young planets that were zipping around the Sun. As this infant planet smashed into the young Earth, it was vaporized along with a good fraction of the Earth in a truly spectacular fireball.

By comparison with this collision, the much more recent asteroid impact that drove the dinosaurs to extinction was a damp firecracker. Assuming that the Mars-sized planet hit Earth at the same speed as the dinosaur-killer asteroid did billions of years later, the kinetic energy released by the impact would have been over a billion times as great!

Some of the hot debris generated from the collision went into close orbit around the traumatized Earth. Before too long, as the fragments in orbit began to coalesce, Earth's new companion began to form. It was startlingly different from the tiny orb that now bestows its beneficent silvery light on present-day lovers (Plates 4 and 5). At the time immediately after its formation, the Moon was probably a mere 25,000 kilometers away, almost fifteen times as close as it is today.

When the young Moon was above the horizon of primitive Earth, it would have extended across a twentieth of the sky. The part illuminated by the Sun would have blazed silver, as the Moon does now, but the part in shadow would have glowed sullenly with its own red light—illuminated by the sudden, white-hot sparks of meteorites as they blazed through the Moon's rapidly dispersing atmosphere and crashed on the lunar surface. And the Moon would have appeared to hurtle through the sky with great speed—Earth was spinning, partly as a result of the great Moon-forming collision, perhaps as much as three times as quickly as it is today.

Since the time it formed, the Moon has receded to its present orbit nearly 400,000 kilometers from Earth. Thanks to the Apollo missions, which left mirrors on the lunar surface that allow the Earth-Moon distance to be measured precisely by lasers, we know that it is still moving away from us at the almost imperceptible rate of three to four centimeters per year. The rate was much faster, however, soon after the Moon's formation. Estimates of how quickly the Moon receded vary, but it could be that at the time life appeared some 4 billion years ago, the Moon was only half as far away from Earth as it is today.

During the early period of its history, when the Moon loomed ominously close, one of the consequences of this intimacy would have been enormous tides. Initially, tidal forces in the Earth-Moon system would have been three hundred times stronger than they are today. Tidal friction distorted the shapes of both Earth and Moon and rapidly dissipated the rotational energy in the system. This slowed Earth's rotation, and it slowed the less massive Moon's even more until it finally turned only one face toward Earth as it does today.

These huge tidal forces would have affected the young Earth in a dramatic way. We are of course familiar with oceanic tides, but we do not realize (because we are standing on it) that tidal deformation of the crust itself takes place twice a day as the Earth rotates relative to the Moon and Sun. This deformation is relatively minor at present, amounting to only about twenty centimeters, but on the early Earth, it was much larger because the Moon was so close and because the freshly formed crust was thinner and more malleable. The crust rose and fell an astonishing sixty meters twice a day. The entire planet must have groaned like a prisoner on the rack under the fierce pull of the red and angry Moon.

Earth's tides soon became less overwhelming as the Moon receded. But they may, as we will see, have played an important role in the formation of life.

A Hellish Young Earth

Even after the trauma from the birth of the Moon had subsided, the early Earth would not have been a nice place to visit. For the first half billion years of its existence, it was being peppered by mountain-sized planetesimals and comets, which astronomers lump together under the term *bolides*. These bolides were the debris left over after the planets had formed. At the beginning, the bombardment from space was so intense that one of these huge objects slammed into Earth on average every thousand years or so, releasing an amount of energy billions of times greater than the nuclear bombs dropped on Japan at the end of World War II.

The growing heat from impacts and from the decay of radioactive elements trapped in the interior of Earth soon melted the planet. Heavy elements such as iron and nickel sank to the center and formed a liquid metal core, while the lighter silicates floated to the surface. The basaltic mantle in between was liquefied as well. Surface temperatures at this time were likely so hot that the Earth was covered with an ocean of molten rock. No wonder geologists refer to this period of Earth's history as the Hadean!

The planetesimals that had earlier coalesced to form Earth were transformed by the raging heat. They had formed through relatively slow impacts during the earliest period of the Solar System's history, and when they crunched together during the first stages of Earth's coalescence, these impacts too were relatively gentle. But now, with Earth so massive, the planetesimal material that made up the Earth was being baked by the heat generated by impacts and radioactive decay. Gases trapped in the original planetesimals, such as hydrogen, ammonia, methane, carbon dioxide, and water vapor, were driven out into the atmosphere. Earth soon became shrouded by hot gases.

This first atmosphere was thick and cloudy, laden with water vapor. Some of the water vapor would eventually condense and contribute to the oceans, but this was far in the future. No liquid water could yet exist on Earth's molten surface.

Earth at this stage must have been a spectacular planet, its thick atmosphere swirling with color as Jupiter's atmosphere does today. The day side would have been dazzling, and the darkness of the night side would have been dispelled periodically by the flashes of huge lightning storms and the fitful, red glow of mighty volcanic eruptions. Occasionally, bright explosions would blaze up out of the haze surrounding the planet, as bolides the size of Mount Everest plunged down to strike the surface.

Similar events were taking place on Mars and Venus as well as on Earth. As the time between major, large impacts lengthened somewhat, a fragile crust was able to form. The mantles of the three planets soon began to churn in huge convection currents like pots of thick, boiling soup. The nature of the currents was very different on the three planets, however. Because Mars was small and its internal heat was not so fierce, convection remained relatively localized. Volcanic upwellings from the mantle formed relatively static "hot spots" that persisted for hundreds of millions of years, resulting in the formation of huge shield volcanoes like Olympus Mons, the largest known volcano in the Solar System (Figure 3.3).

THE SPARK OF LIFE

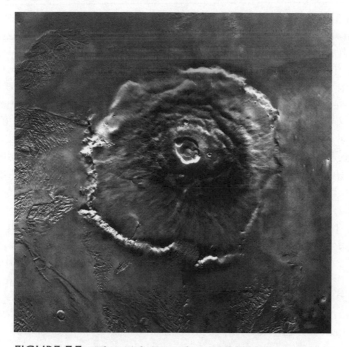

FIGURE 3.3 The mightiest volcano in the Solar System, the great shield volcano of Olympus Mons on Mars. This Martian volcano is more than three times taller than Mount Everest and wider than the entire Hawaiian Island chain. Its top, however, is almost completely flat. (Courtesy NASA)

On Venus, the crust was so thin and fragile that convective mantle currents periodically swept most of it away, like the thin layer that forms and breaks up on the surface of a boiling sugar solution. Indeed, this process is still going on. The last replacement of the old surface of Venus with a new one, which wiped most of the planet clean of accumulated craters and other features, took place a mere half billion years ago.

Earth lay between these extremes. Great convective cells formed in the mantle, as cooler material plunged down, became heated, and rose back up. But the crust was thicker and more stable than on Venus, and gradually a pattern emerged in which pieces of this crust began to move about the surface, carried by the mantle's convective motion. This was the beginning of plate tectonics, which have shaped continents and mountain ranges and have played a greater role in the history of Earth than in those of any other planet or moon in the Solar System.

All this time, planetesimals and comets continued to hurtle toward Earth, though not at quite the same devastating rate as during its earliest history. Nevertheless, the infall of these giant bodies was so common that the energy generated by their impacts far exceeded the energy coming from the faint, young Sun. These impacts probably blasted away a good fraction of Earth's first, and strongly reducing, atmosphere. However, the bolides themselves contained gaseous components and water, and this helped replace some of the young Earth's atmosphere as it was being lost.

Today, water-rich comets are plentiful in the outer reaches of our Solar System. It has been suggested that such comets were the chief sources of the water that would eventually accumulate on the Earth's surface and form its extensive oceans. But it has now been found that levels of deuterium (an isotope of hydrogen) are twice as high in the water of the comets Halley, Hyakutake, and Hale-Bopp as they are in the Earth's seawater. This means that comets could not have been the only source of the water in the oceans. The Earth's water seems to have come primarily from hydrated minerals making up part of the planetesimals that had accumulated to form the planet.

Certain types of meteorites contain significant amounts of organic materials, and comets carry a rich assortment of simple organic molecules. Unfortunately for the possibility of a primeval soup, most of these extraterrestrial organic compounds were probably destroyed on impact. Even if some of them survived, they would have been fried to gaseous breakdown products on the Earth's hot surface. Nonetheless, these decomposition products would have contributed important gases to the Earth's atmosphere.

Volcanic eruptions, however, were probably the most important source of gases to replenish the Earth's early atmosphere (Plate 6), and volcanoes played important roles on other planets and moons as well. Today, the most volcanically active body in the Solar System is Io, the innermost moon of Jupiter. Recent images of Io sent back to the Earth by the Galileo flyby mission have revealed intense volcanic hot spots where lava temperatures exceed 1,300°C, hot enough to melt essentially any type of rock. Plumes of gas rise up above Io's surface to ten times the height of Mount Everest (Figure 3.4). Io's volcanoes are driven by tidal heating caused by the overwhelming presence of nearby Jupiter. In

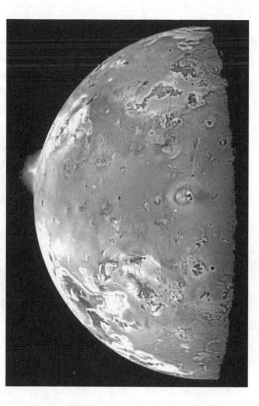

FIGURE 3.4
Two volcanoes erupting simultaneously on Io. The one near the moon's edge is producing a plume 140 kilometers high. The one in the center of the picture has apparently been erupting continuously for the last eighteen years—it has been detected in Voyager images sent back to Earth in 1979. (Courtesy NASA)

the early Solar System, the volcanic activity on Mars, Venus, and our own planet would have rivaled that of present-day Io.

In spite of all these sources that tended to replenish it, the Earth's atmosphere must have undergone great changes early in its history. The most important was the loss of its original hydrogen, which was so light that it could escape and disappear into space. As hydrogen became scarcer and scarcer in the atmosphere, the reduced gases methane and ammonia that remained would have become unstable and converted into carbon dioxide and nitrogen. This process would have been accelerated by the fierce short-wave ultraviolet light that bathed the Earth.

It is this process that worried Philip Abelson. If methane and ammonia had been destroyed completely, there would have been little chance that syntheses of amino acids and other compounds could have taken place on the early Earth as it cooled. There would have been no ingredients for the primordial soup.

THE EARTH'S APOCALYPTIC BEGINNINGS

The Abelson dilemma may not be as problematic as it seems, however. Even after most of the Earth's hydrogen had been lost, volcanoes might have still supplied and helped maintain reducing gases in the atmosphere. If a steady state were reached between volcanic release of reducing gases and their destruction by ultraviolet light, this may have permitted significant amounts of methane and ammonia to persist in the Earth's primitive atmosphere.

Another protecting factor might have been the Earth's atmosphere, which may have been so opaque that the opacity itself protected against the destructive effects of ultraviolet light. Chris Chyba, now at the SETI Institute in Mountain View, California, along with the late Carl Sagan of Cornell University, suggested that the early atmosphere may have been even smoggier than the thick, yellow haze that enshrouds modern-day Mexico City. The Earth's surface may have been perpetually dark, because sunlight could not penetrate.

In the topsy-turvy environment of the early Earth, this witches' brew of chemicals would actually have protected methane and ammonia from destruction. Rather than dealing death and illness as smog does today, it would have helped to preserve organic molecules as they were being synthesized and would have helped to make the beginnings of life possible.

Eventually, however, as hydrogen was lost to space and the input from volcanic discharges diminished in intensity, only trace amounts of methane and ammonia would have remained. At this point, the atmospheric gases would have consisted of carbon dioxide, nitrogen, and water vapor. Exactly when this transition took place is not known, for this whole period is missing from the rock record. But the longer the change was delayed, the greater the quantities of organic compounds that could have been synthesized on the young Earth.

The Transition to a Habitable World

During this early, tumultuous Hadean period of the Earth's history, conditions were so severe that life could not have begun. However, after several hundred million years of earthly hell, things began to change.

There are now no remnants of the thousands of craters that must have pockmarked the early Earth's surface. Nonetheless, by simply looking at the Moon, we can see that the frequency of impacts must have been declining during the Hadean period. Much of the Moon's early surface is still intact, because it was not destroyed by plate tectonics, and as a result many more early craters have survived. The pattern and ages of these craters show that by about 3.8 to 3.6 billion years ago, the frequency of large impact events had diminished to something like the present rate of one large ten-kilometer bolide impact event every hundred million years.

Around 300 to 400 million years after the Earth had accreted to its present size, the impact frequency had declined to the point that the planet's surface temperature started to drop and a thick crust could solidify. Finally, as the crust cooled below the boiling point of water, the oceans began to form. The boiling point was probably substantially higher than the current sea-level boiling point, because the atmosphere was much denser. Thus, water oceans could actually have begun to form at temperatures in excess of 100°C.

On the Earth today, enormous quantities of water are discharged into the oceans from rivers. The flow is so great that it would take only 40,000 years to fill up the modern ocean basins. But the original formation of the oceans must have taken much longer. At first, evanescent ponds would have appeared briefly on the coolest parts of the crust. They would have vanished again as the season changed or as the rocks beneath them were episodically heated. But eventually, probably over a span of a few million years, stable oceans formed. Although some of the water on the early Earth could have been blasted into outer space by bolide impacts, the amount of water has probably not changed much since the early ocean first formed. This first ocean probably had a volume similar to that of today's oceans.

Life as we know it could not have appeared without bodies of water. Because of Schopf's work, we know that life was well established on the planet by 3.5 billion years ago. Thus the Earth's oceans must have originated roughly sometime between 4.2 and, at the latest, 3.6 billion years ago.

Even after the Earth's oceans formed, however, the threat of impacts would still have made the appearance of life very difficult. At least ten large bolides struck the Moon between about 3.8 and 4.1 billion years ago. The great "seas" that are such striking features of the lunar surface today are apparently the result of a massive bombardment that took place during this period. Whether this was the tail end of a continuous bombardment that had gone on for the previous half billion years, or whether it was the last of a series of periodic bursts of impacts that each lasted for a relatively short time, is still being argued.

Scientists who belong to the periodic-impact school suggest that these events were caused by occasional disturbances of the great cloud of comets that surrounds the Solar System. These disturbances, perhaps caused by passing stars, sent volleys of comets spiraling inward toward the Sun, and a few of them slammed into the Earth and the Moon.

Impact events of such magnitude on the Earth would have released enough energy to vaporize the oceans and sterilize the planet in a burst of superheated steam. Kevin Maher and David Stevenson of the California Institute of Technology have suggested that life could have in fact originated several times but that it was destroyed again and again by these violent, sterilizing impacts. Perhaps life might have taken a very different form, were it not for these repeated "impact frustrations."

Prior to about 3.5 billion years ago, much of the land exposed above the ocean surface on the early Earth probably consisted of volcanic islands scattered about over particularly thin parts of the crust. Unlike the great volcanoes of Mars, which grew to enormous size at around the same time because the Martian crust was not moving, each of these volcanoes stopped growing as soon as the hot spots moved away from underneath them.

The hot spots would have left behind chains of islands, the youngest near or directly over the hot spot and the remainder of the islands growing progressively older the further they were from the hot spot. Today's Hawaiian islands give us a glimpse of what those early island chains might have been like. A hot spot beneath the floor of the Pacific has generated the Hawaiian chain as the crust has moved over it. Each island of the chain lasts for about fifty million years until the sea wears it away. Nubs of old Hawaiian islands, barely breaking the surface in the shape of a few scattered reefs, stretch in an unbroken chain across the Pacific all the way to Midway Island two thousand kilometers to the west. Then the chain veers north and stretches thousands of kilometers further. Each of these islands was once as large as the present-day Big Island of Hawaii.

The first islands to form in the early ocean would have been very different from the earthly paradise currently enjoyed by visitors to Hawaii (Plate 7). With no living coral reefs to protect them, the volcanic islands would soon have been weathered away by storms and reduced to roiling areas of foam and jagged, black basalt. Even the few granitic islands, the seeds of future continents, would have been subjected to strong erosional forces.

One component of the early volcanic discharges would have been hydrogen chloride gas, which dissolves in water to form hydrochloric acid—the same acid used to keep swimming pools clean. The acidic, primitive oceans would have rapidly eaten away acid-soluble minerals, accelerating the erosion process still further. Erosion of these early islands would soon have produced the salty ocean we are familiar with today.

Last but not least were the mighty tides. Because the Moon was still close, the oceanic tides, even in the middle of the newly formed ocean and far from any land, would have risen and fallen several meters, compared with three-quarters of a meter today. On islands that happened to stand athwart these massive movements of water, the tides could easily have swept across far wider regions than would be accessible to the tides of today. The enormous tidal floods, surging up and down the islands' black sand beaches and mudflats every day in predictable patterns, must have played an important role in the synthesis and sorting of organic compounds on the early Earth.

A Boiling Scylla or a Frozen Charybdis?

As the hammering from asteroids and comets gradually diminished, the major source of energy at the Earth's surface would now have been provided by sunlight. One might think that the Sun blazed in the sky— it was, after all, a young star. But it turns out, surprisingly, that the Sun of four billion years ago was only about three-quarters as luminous as it is today. This means that if conditions on the Earth had been the same as they are now, the entire surface of the planet would have been frozen. The Earth would have resembled Jupiter's ice-covered moon, Europa.

There is another possibility, however. In spite of the dim and chilly Sun that hung in the sky, the surface of the early Earth might actually have been warm, even ovenlike. This is because of the so-called greenhouse effect.

In a greenhouse, the walls and roof are made of glass or some other transparent material. Sunlight can thus penetrate and be absorbed and converted to heat by the plants and other objects in the greenhouse. This warms the inside of the greenhouse and produces the balmy conditions that permit plants to grow even in winter.

If the atmosphere were completely transparent to infrared radiation, the infrared light reflected off the surface of the Earth would be lost into space. But just as the glass of a greenhouse traps heat, gases such as carbon dioxide, methane, and ammonia trap some of the outgoing infrared radiation. This heats the gases, making the whole atmosphere warm up.

The most important of these greenhouse gases at present is carbon dioxide. On a per-molecule basis, however, methane is twenty-one times as effective a greenhouse gas as carbon dioxide. Even a little methane goes a long way. There is not much methane in the air we breathe, less than two parts per million, but it is rapidly growing in concentration—from, among other sources, the flatulence of the vast herds of cattle that we raise. Methane is so opaque to infrared that even these small amounts account for about 10 percent of the total greenhouse effect that is leading to global warming.

Moderate levels of methane, ammonia, and carbon dioxide could have kept the Earth warm during its early history. But there is a delicate balance here. Too little of these gases, and the Earth would have been covered with ice. Too much, and it would have turned into an oven. Before life appeared, there would have been no moderating effect of living organisms that might have prevented either of these extreme conditions from developing.

It is amazing in retrospect that the Earth was neither too hot nor too cold for life to appear. Indeed, there is good evidence that the Earth nearly did freeze over twice, even after life had appeared. It was largely covered by ice and snow during two massive glaciation events, one 2.2 billion years ago and one about 800 to 500 million years ago. Apparently, volcanic activity eventually pumped enough carbon dioxide into the atmosphere to cause these glaciers to melt.

Even during earlier times, global glaciations probably occurred. But if the early Earth did sometimes freeze over, how was it eventually defrosted, and what did these periods of extreme cold do to emerging life? The jury is still out on these questions.

We can see the unmoderated effects of atmosphere composition when we look at our neighboring planets. Svante Arrhenius was the first to calculate that if Venus were to have the same atmospheric composition as Earth's, its temperature would be only about an average of 35˚C. We could live there, for although it would be blazing hot at the equator, it would be bearable at the poles.

But we now know that the atmosphere of Venus contains around one hundred times as much carbon dioxide as the atmosphere of Earth. As a result, its surface temperature is a roasting 500°C. Interestingly,

Venus and Earth have the same amount of total carbon dioxide. On Earth, however, most of it is sequestered away in the form of carbonate rocks, whereas on Venus it has all been baked out into the atmosphere.

At the other extreme, even though the atmosphere of Mars consists almost entirely of carbon dioxide, it is so thin that any greenhouse warming has little effect. As a result, the Martian climate is freezing cold.

The fates of our sister planets are dramatic cautionary tales. They demonstrate that before life appeared on the early Earth, our planet was not only lifeless but could easily have remained lifeless forever. Its climate might have tipped permanently in the direction of an icebox or an oven. It is fortunate for the emergence of life that this did not happen, but we are still uncertain about why it did not. And indeed, even after life arose, what would have prevented the entire Earth from suddenly shifting toward a climatic state in which these first living creatures would have been snuffed out?

James Lovelock in England and his American collaborator Lynn Margulis at the University of Massachusetts have speculated that feedback processes generated by living organisms have kept the Earth's temperature within a narrow range tolerable for life throughout much of its history. Novelist William Golding named Lovelock's idea the Gaia hypothesis, after the Greek goddess of the Earth. The hypothesis suggests that life optimizes conditions on the planet.

To say the least, this idea received a very unenthusiastic reception at first from other scientists. Lovelock's initial suggestion was that all the living things on the Earth are linked together into a single living super-organism, which regulates conditions on the planet for its own survival. As Lovelock presented it, the Gaia concept sounded uncomfortably like some new-age Earth Mother religion—and it was immediately embraced with enthusiasm by some very unscientific groups indeed.

More recent versions of the Gaia hypothesis have been far less extreme and rely on known biological and chemical processes. For example, living organisms should be able to help regulate the amount of carbon dioxide in the atmosphere within fairly narrow limits. Here is how it seems to work (Figure 3.5):

As solar luminosity increases over time, the temperature rises. Living organisms are then able to multiply in greater numbers. When they die, their remains accumulate as carbon-rich sediments, such as the massive calcium carbonate deposits formed today from corals, seashells, and

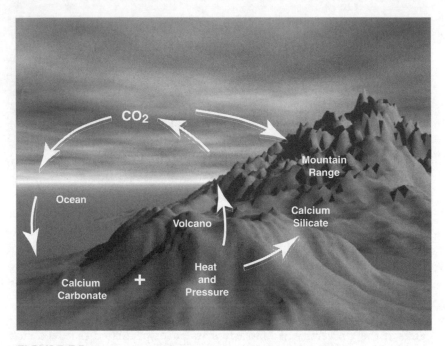

FIGURE 3.5 Volcanoes and plate tectonics play essential roles in regulating the concentration of carbon dioxide in the atmosphere and our planet's overall temperature. Carbon dioxide dissolving in the oceans precipitates as calcium carbonate, reducing the amounts in the atmosphere. Then heat and pressure deep in the Earth change the calcium carbonate into calcium silicate, and some of the resulting carbon dioxide gas is released by volcanoes. Finally, as mountain ranges are pushed up by continental drift, the silicate is exposed and weathered, which removes carbon dioxide from the atmosphere again. Similar cycles were operating even before life appeared, but life has helped to moderate the earlier, violent swings in carbon dioxide concentration.

other marine organisms. The carbon-rich sediments accumulate at the bottom of the sea, thus acting to remove carbon from the system. Then, as the amount of carbon dioxide in the atmosphere is reduced and the world begins to cool again, this reduces the level of the photosynthesis that depends on it. The oceans can support less life and fewer carbon-rich sediments are sequestered.

All the time that this is going on, the carbon-rich sediments that are buried deep within the Earth are metamorphosed by heat and pressure into calcium silicate, and some of the resulting carbon dioxide escapes back into the atmosphere. This tends to raise the temperature again. But the impact of this carbon dioxide release is tempered by the transfer of massive beds of metamorphosed calcium silicate to the continents through mountain building, where it cycles back to the surface and is exposed to weathering. The calcium silicate is converted back to calcium carbonate, and carbon dioxide is removed once more. Thus, both living organisms and geological processes play a role in temperature regulation.

Without the presence of living organisms, this process might spin out of control. Even with the moderating influence of life, the cycle has huge delays built into it, which can have dramatic consequences. For example, silicate-rich metamorphosed ocean sediments have been uplifted to form the Tibetan plateau and the Himalayas during the last five million years. Maureen Raymo of the Massachusetts Institute of Technology has pointed out that these new mountains began to erode as soon as their silicate beds were exposed to the air. The rocks absorbed carbon dioxide in huge quantities during weathering, reconverting the calcium silicate to calcium carbonate. She thinks that it is not a coincidence that the recent series of ice ages also began about five million years ago.

When presented in this way to skeptical scientists, the Gaia hypothesis makes a great deal more sense than it did in its first mystical manifestation. Variations on the Gaia theme may account for the Earth's apparent ability to keep to a narrow range of temperatures as the Sun has gradually warmed over the last four billion years.

But what about the time before the appearance of life? Chemical versions of these cyclic processes took place on the early Earth, but without the Gaia-like feedback provided by living organisms. Carbon-containing rocks must have been recycled by the movement of tectonic plates and the gradual formation of the continents, for otherwise the process would have spun out of control and all the planet's carbon would soon have become inaccessible. Tectonic recycling must have been essential to the survival of the Earth as a potential abode for life, just as the absence of continuing tectonics doomed Mars and may have doomed Venus.

Scientists would cheerfully sell their souls to get a glimpse of the missing first 700 million years of the geological record. This would tell us a great deal about the true conditions on the early Earth, and perhaps settle the question of how it managed to steer between a boiling Scylla and a frozen Charybdis.

THE EARTH'S APOCALYPTIC BEGINNINGS

All we can tell at the moment is that during much of that early time, the planet would not have been able to support life. Even as the early Earth was becoming more and more quiescent, it was still lifeless. Appreciable amounts of organic compounds were probably not present when the bolide bombardment finally slackened, because they would have been destroyed during the Earth's earlier, violent past. To set the stage for the origin of life, some sort of prebiotic soup must have been generated. Was it cooked at home? Or did the ingredients come from space? What was it like? Was it thick and strong, or was it—like the soups cooked by well-meaning Victorian ladies for the poor—too watery to have much nourishment?

4

PREBIOTIC SOUP: THE RECIPE

[W]e conclude that the porous mantle of the earth thus supplied by planetesimal infall with unstable carbides, nitrides, phosphides and sulphides undergoing transformation into more stable compounds, and generating during this process hydrocarbons, ammonia, hydrogen phosphide and hydrogen sulphide gases mingled with the ordinary gases carried by the planetesimals, furnished rather remarkable conditions for interactions and combinations, among which unusual synthesis would not be improbable.

Thomas C. and Rollin T. Chamberlin, Science 28 (1908): 904

IN THE LATE AFTERNOON OF MARCH 15, 1806, AT ABOUT 5:30, loud explosions shook the area near the French town of Alais. Two large stones hurtled from the sky, one of them hitting a branch of an olive tree. Chemists analyzed the stones shortly afterward and found that they contained a surprising amount of carbon, constituting several percent of their total weight. These meteoric stones could not have acquired this material from terrestrial sources, for they had been picked up soon after they fell.

During the decades that followed, analyses of stones collected from other meteorite falls, especially the Orgueil meteorite that fell in France in 1864, confirmed the presence of organic compounds. The controversy about the origins of this unusual class of meteorites, now called carbonaceous chondrites, has continued almost to the present. It reached a peak in the 1960s, when Bartholomew Nagy and his coworkers published microscopic investigations and state-of-the-art analyses of the Orgueil meteorite and other carbonaceous chondrites. They found tiny round shapes, which looked to them as if they might be fossilized algae. Chemical analysis revealed oil-like hydrocarbons. Based on their similarity to those found in sediments on the Earth, it was suggested that the hydrocarbons were of biotic origin.

These claims generated a great row and an intense, often nasty debate. Most of the objections were centered on the omnipresent problem of terrestrial contamination, especially since the meteorites had sat on museum shelves for long periods. The Orgueil hydrocarbons were considered to be terrestrial contaminants, probably introduced into the meteorite by airborne dust and smoke particles. Several investigators eventually concluded that the "organized elements" were nothing more than ragweed pollen.

The search for amino acids in meteorites, utilizing modern analytical techniques, also began in the early 1960s and confirmed that these key biomolecules were indeed present in several carbonaceous chondrites. However, in many cases the distribution of the observed amino acids was disturbingly similar to that of human fingerprints. As one analyst, P. B. Hamilton, put it, "What appears to be the pitter-patter of heavenly feet is probably instead the print of an earthly thumb."

Clearly, the problem of terrestrial contamination could only be avoided if fresh, pristine extraterrestrial samples could be obtained. And in the late 1960s, this was going to happen—the Apollo missions were about to return samples from the Moon.

No one was sure what to expect. Although amino acids were of paramount interest, the scientific community realized that merely detecting amino acids in lunar samples would not be likely to resolve the question of their origin. Something else was needed, and that something was a way to investigate the optical activity of these molecules.

As Pasteur had shown over a century earlier, the amino acids derived from living organisms are different from those synthesized in the laboratory (see Chapter 1). Biological amino acids consist of only the left-handed isomers (L-isomers), whereas synthetic amino acids are racemic mixtures, in which the amounts of the left- and right-handed (L- and D-isomer) forms are equal and their tendencies to rotate light cancel each other out (Figure 4.1).

If amino acids in lunar samples and meteorites were made by natural abiotic reactions of the type mimicked in the Miller-Urey experiment, they should be present as a racemic mixture. If they were contaminants, they would consist only of the L-isomers. Of course, if they were from organisms living on the Moon, they would probably not be racemic. If, for example, lunar life had arisen independently of life on Earth, its amino acids might consist entirely of the D-isomers.

By the time the first lunar samples collected during the Apollo 11 mission arrived back on the Earth at the end of July 1969, methods for the

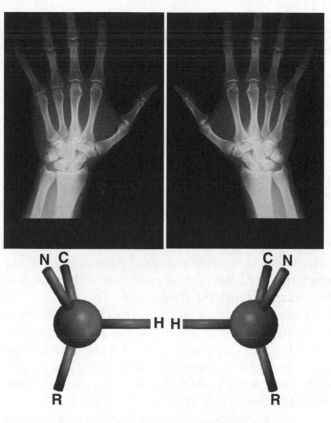

FIGURE 4.1   The architecture of left-handed (L-isomer)
and right-handed (D-isomer) amino acids is similar to that
of our own left and right hands. The amino acid chemical
formulas are identical, but the various components are
joined to the carbon atom in the center (represented by
the central ball) in different ways, producing molecules
that are mirror images of each other. C is the acidic, or
carboxyl, group; N is the basic, or amino, group; and R is
the side-chain group. All the protein amino acids have a
hydrogen atom (H) at the central carbon (glycine has two
and is therefore not optically active). However, many
amino acids found in meteorites have another R group
instead of a hydrogen atom.

routine separation of the D- and L-isomers of amino acids had been developed. There was great excitement and high expectations during the
initial analyses, some of which were carried out in the laboratory of

Cyril Ponnamperuma at the NASA Ames Research Center in Menlo Park, California.

The results were extremely disappointing. If there were amino acids in the Apollo lunar samples, they were present in amounts statistically indistinguishable from those seen in blanks carried through the same analytical procedure.

A Message from Outer Space

Luckily at this point, fate intervened, and in a totally fortuitous guise. At about 11:00 A.M. (local time), September 28, 1969, a meteorite fell near the town of Murchison, Victoria State, in southern Australia. The parent object broke up during flight and scattered many fragments over an area of about seven square kilometers, with many pieces breaking further upon impact. Most of the surface of the individual pieces was covered by a thin, melted fusion crust produced by the intense heat as the fragments blazed through the atmosphere.

Several stones were picked up soon after the fall and sent to various laboratories for examination. It was soon recognized that the meteorite was a carbonaceous chondrite, the type known to contain organic material. This was the first time that a truly fresh, uncontaminated meteorite was available for study, at a time when all the newest chemical tricks could be brought to bear on it.

When a piece of Murchison arrived in Ponnamperuma's laboratory in November 1969, he gave it to geochemist Keith Kvenvolden, the scientist responsible for many details of the lunar analyses, and told him, "Please analyze this." Kvenvolden, along with a team of investigators assembled by Ponnamperuma, embarked on the job with great caution.

Excitement was high. The excellent condition of the recovered stones and the availability of state-of-the-art laboratory procedures offered for the first time an opportunity to discover exactly what compounds might be present in a pristine meteorite.

The team was determined to avoid the mistakes and wild claims of the earlier analyses. But as the list of organic compounds that they were finding grew, Kvenvolden and his co-workers realized that they were glimpsing an unimaginably remote world in both time and space. And that world, in spite of its foreignness, had a familiar feel about it.

They were detecting amino acids and other compounds that bore an uncanny resemblance to those that Stanley Miller had produced in his experiments. The amino acids could not be terrestrial contaminants,

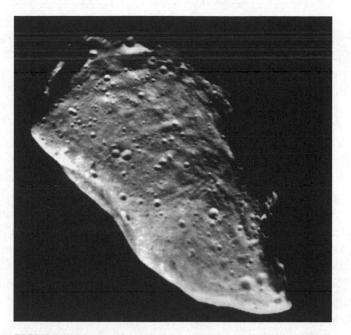

FIGURE 4.2 Our best look at an asteroid so far. The picture shows the asteroid 951 Gaspra and was taken by the *Galileo* spacecraft on its way to Jupiter. The asteroid measures about nineteen kilometers in its longest dimension and shows signs of numerous impacts. Analysis of the Murchison meteorite shows that Miller-Urey-type reactions took place in the depths of such asteroids as the Solar System was forming. (Courtesy NASA)

because they were racemic mixtures, made up of equal quantities of the D- and L-isomers. Apparently the Murchison meteorite had come from an asteroid that had contained an early Solar System primordial soup (Figure 4.2). Somewhere in the far reaches of space, about 4.5 billion years ago, it appeared that nature had carried out an experiment almost identical to Miller's.

Kvenvolden was greatly excited by his results, but he was soon in for a nasty surprise. While the Murchison analyses were in progress, Ponnamperuma had set off for Iceland to explore hot springs for evidence of life. Soon after his arrival, he slipped into a pool of boiling water and severely burned his leg. He was hospitalized at the Stanford Medical Center when Kvenvolden visited to brief him on the results: "I have

PREBIOTIC SOUP: THE RECIPE

some really good news. The amino acids in Murchison are racemic." Ponnamperuma responded, in a fit of envy, "You are no longer responsible for this project. And don't tell anyone about these results."

Kvenvolden was stunned. He and the team had found racemic amino acids in Murchison, which demonstrated that they were abiotic in origin. Somehow, these amino acids had been made by natural abiotic processes. These compounds must have been made on the Solar System body, whatever it was, that Murchison had come from. Was he to be excluded from this pivotal discovery?

After passionate argument, Ponnamperuma relented and a paper describing the Murchison results was submitted to the prestigious journal *Nature,* with Kvenvolden as first author. This did not prevent Ponnamperuma from continuing to try to capture the limelight, however. He scheduled a press conference a month before publication of the paper and failed to invite Kvenvolden or any of the other co-authors of the paper. Kvenvolden found out about this the day before and complained directly to the head of the research center, who replied, "Don't worry. Just show up at the time of the scheduled press conference."

The next day, official places were given to all the paper's authors. Ponnamperuma, who had become progressively more autocratic in his role as chief of the lab, was furious. But the story was now out, and its implications were enormous and continue to reverberate. The primordial soup was real, not simply some construct of imaginative chemists. Haldane and Oparin, Miller and Urey had all been on the right track after all.

How Many Tureens of Soup?

Careful analysis has now revealed that the Murchison meteorite contains a cornucopia of unusual compounds. John Cronin and Sandra Pizzarello at Arizona State University have identified over seventy different amino acids, most of which are unique to the meteorite and do not occur naturally on the Earth. And remarkably, along with the amino acids, adenine, and some of the other nucleobases that are found in DNA and RNA have been found.

All the evidence indicates that a Miller-Urey type of synthesis had taken place on the Murchison meteorite parent body. And there were other resemblances to the Miller-Urey experiment as well. Numerous as the amino acids and other small compounds were, they constituted only a small percentage of the total organic material present in the meteorite. An astonishing 2.5 percent of the meteorite was organic carbon, consisting primarily of the same sort of polymeric, oily goo that had coated the inside of Miller's apparatus after he had sparked his gas mixture.

Murchison is not unique. Other carbonaceous meteorites, including some collected during the last couple of decades in the ice fields of Antarctica, have also been found to contain a similar complex mixture of organic compounds.

How were these compounds produced? It appears that conditions enabling Miller-Urey syntheses to take place existed on or in some of the earliest planetesimals as they condensed out of the solar nebula during the formation of the Solar System. The planetesimals probably had no atmosphere, since they were so small. But perhaps they contained internal cavities that could have trapped gases and reagents such as hydrogen cyanide, aldehydes, ketones, and ammonia, along with ice.

As radioactive decay generated heat in the planetesimals' interiors, liquid water was released and dissolved the reagents. Just as in the water flask in Miller's apparatus, Strecker-type reactions would have then taken place in the planetesimal water solution, aided by the energy produced from the radioactivity and producing a *Beilstein*-like mixture of organic compounds (see Chapter 2). Eventually the water would have diffused to the surface and been lost to space, leaving behind the organic materials. Bits of these organic-laden planetesimals, blasted off by subsequent collisions with other Solar System bodies, have been landing on the Earth ever since.

Remarkably, the molecules carried by the recent arrivals such as Murchison have survived the ravages of time and the radiation flux of outer space for nearly 4.5 billion years. During this immense span of time, some of the organic compounds may have broken down. Was the mixture of organic compounds on the original planetesimals far richer than the traces that chemists have found in the Murchison meteorite? We may find answers to this if space probes can be designed to penetrate to the protected hearts of asteroids, where richer collections of organic molecules might have been preserved.

The source of Murchison was almost certainly the asteroid belt. Many asteroids, such as 130 Elektra, have been shown by optical remote sensing methods to contain organic substances. There are numerous asteroids and small moons of the outer Solar System that seem to resemble carbonaceous chondrites. Only about 1 percent of the meteorites that land on the Earth are of this type, but this is probably not a measure of their true abundance. Carbonaceous chondrites are fragile and tend to break up in the atmosphere—the Murchison fragments themselves were bits of a much larger body, most of which exploded into tiny chunks that burned away during their fiery descent. And the surviving bits of such meteorites tend to weather away rapidly once on the Earth. We are seeing only a small fraction of these carbon-rich bodies.

Although Kvenvolden's team had shown that there must have been a tureen of primordial soup on (or more probably deep inside) the parent asteroid of the Murchison meteorite, a larger question remained. Could a primordial soup have been cooked on the Earth as well?

If the Earth had a reducing atmosphere that persisted after the last great bolide impacts, then there is no doubt that amino acids and other organic compounds would have been synthesized easily. As we mentioned earlier, the key components needed for direct prebiotic synthesis may have come directly from volcanoes belching the needed gases. They may have also been created during the blazing collisions of comets and asteroids with the Earth.

But as we have also seen, it is possible that the early atmosphere did not contain methane, ammonia, and hydrogen, but rather a more chemically neutral mixture of carbon dioxide and nitrogen. This gives rise to Abelson's dilemma—if a carbon dioxide/nitrogen mixture is used in a Miller-Urey experiment, few or no simple organic compounds are synthesized.

The direct home-grown synthesis may not have been the only way in which the Earth's prebiotic soup was prepared, however. Perhaps some of the soup's components could have come directly from space. How important could such a process be? In part it depends on how plentiful these building blocks were—and are—elsewhere in the Solar System.

Life's Seeds from Space?

Even after it became a potentially habitable world, the early Earth was still accumulating a great deal of debris from space. Along with the occasional large comet and asteroid impact, numerous meteorites, some the size of office buildings, often blazed down out of the sky. Many of them probably exploded in the atmosphere rather than directly hitting the surface. This is still going on today and is apparently what happened in June 1908 in Siberia, when the Tunguska meteorite exploded with a blast like a nuclear bomb and knocked down trees over an area equivalent to a large, modern city.

Some meteorites that fell on the early Earth were certainly carbonaceous chondrites. Perhaps these meteorites seeded the Earth with the organic ingredients needed to make a prebiotic soup. In a sense, this is the modern-day version of panspermia, although what was arriving on the planet would not have been life itself but rather the building blocks of life.

Could this extraterrestrial material produce a soup rich enough to support the origin of life? The answer is very different depending on whom you ask. Stanley Miller says that extraterrestrial material would not have been a significant component of the primordial soup. He points out that the amounts of any organic compounds coming from space would have been dwarfed by those that could have been made had there been a reducing atmosphere, even if those reducing conditions were episodic and transient. Carl Sagan, on the other hand, suggested that because we do not know whether a few key compounds present in the early ocean might have been enough for the origin of life, even the tiny amounts from space could have been sufficient. His colleague Chris Chyba asserts that the delivery of organic compounds from space would have been the only game in town.

Proponents of life's seeds from outer space are faced with the problem that the amounts of organic materials that could have arrived on the Earth must have been very small. Consider for example the infall of Murchison-like material. At present, only about 200 kilograms of carbonaceous meteorites fall to the Earth each year. Most of this material is, of course, simply rock, and only 1 or 2 percent of it consists of organic material. At this rate, the total amount of organic material that would have accumulated over the entire age of the Earth would be puny indeed—in the range of ten-billionths of a gram of organic material per liter of ocean water! Moreover, only a small portion of this would have been delivered to the early Earth during the period when it was needed for the origin of life.

Of course, much larger objects have hit the Earth in the past. The mountain-sized object that collided with our planet 65 million years ago and caused the demise of the dinosaurs weighed 10^{18} grams, or a trillion tons. Surely, if some of the large impactors contained significant amounts of organic carbon, as we suspect that many asteroids and comets do, they could have contributed substantial amounts to the inventory of organic compounds on the early Earth.

Unfortunately for this idea, large-body impacts on the Earth are extremely violent events. Although most organic compounds were probably destroyed during a giant bolide impact, their decomposition products, such as hydrogen cyanide, aldehydes, and ketones, might have survived. These could then have served as raw materials for later prebiotic synthesis reactions on the Earth itself.

Indeed, there is evidence that small quantities of intact organic molecules can survive large impact events. A layer of material (called the Cretaceous/Tertiary, or K/T, boundary) was thrown out across the entire surface of the Earth by the asteroid impact of 65 million years ago. Careful analysis of this layer has turned up traces of some of the same unusual nonterrestrial amino acids that Kvenvolden and his colleagues had detected in the Murchison meteorite. Amino acids were either present in the impacting object, so that some tiny portion of them survived the impact event, or they were actually made during impact, by a process called shock synthesis. Unfortunately, the quantities of amino acids found in the K/T boundary sediments are very, very small. Even if all these amino acids had ended up in the oceans, they would have been diluted to almost nothing.

One promising source of organic materials from outer space remains to be considered: the very small bits of *cosmic dust,* or micrometeorite particles, that abound in space. The Earth today accumulates about 40 billion grams of cosmic dust each year, and even more of this material must have been swept up by the young Earth. Because these particles are so small, they are slowed gradually from their high speeds by the wispy upper atmosphere. They then drift down to the surface almost unmodified by their entry, a cosmic gentle rain from heaven reminiscent of the rain of living spores imagined by Svante Arrhenius. Samples of this dust have been collected by a spacecraft mission, by aircraft flying at high altitudes, and by filtering the water obtained by melting tons of Antarctic ice. These tiny flecks of material tend to be even richer in carbon than carbonaceous meteorites like Murchison.

Alas, when the calculations are done, even all that space dust would have produced an amino acid concentration of only a millionth of a gram per liter in the early oceans. Thus, the good news is that some intact organic compounds must certainly have been delivered to the Earth from outer space. The bad news is that if they were the only source of organic compounds, the result would have been a weak, thin soup that would have shamed even those Victorian ladies.

If the early atmosphere was not reducing, and organic material from outer space was only a trickle, where might a sufficiently thick and nourishing soup have come from? Because this question remains unresolved, some scientists have taken a different tack altogether. They are searching for a source of the prebiotic soup in less obvious places—

within oil slicks on the early ocean or at the sites of hot hydrothermal vent discharges.

A Primordial Oil Slick

The *Exxon Valdez* disaster seems an unlikely event from which to learn lessons about the origin of life. And yet the Alaskan landscape after that disaster may have had a lot in common with the landscape of the early Earth.

Regardless of whether the first organic materials came from outer space or were home-grown, the most abundant component must have been tarry material like most of the organic material in Miller's experiment and in the Murchison meteorite. Unlike amino acids, which are readily soluble in seawater, this material would have been very insoluble. It would have tended to coagulate into gooey lumps or films—like the black, tarry stuff that still persisted a few inches beneath the surface of the beaches of Prince William Sound more than a decade after the accident of 1989.

Indeed, the early Earth may well have been one giant *Exxon Valdez* disaster. Heinrich Holland of Harvard has suggested that the Earth's surface might have been covered by a primordial oil slick, perhaps several meters thick. This slick could have coated the small islands that were present, as well as the surface of the entire world-ocean. It would certainly have dwarfed the mere 11 million gallons of crude oil released during the *Exxon Valdez* spill!

The oily slick could have been a source of organic compounds needed to make a primordial soup. Stanley Miller and his student Eduardo Borquez found in the late 1990s that the brownish, oily tar generated in prebiotic experiments slowly releases substantial amounts of adenine when exposed to water. Amino acids have also been found to be released slowly from such tars. Thus, the primordial oil slick could have been like a giant time-release capsule, continuously supplying adenine and amino acids to the oceans of the early Earth.

The oily stuff could also have filled some other gaps in the prebiotic chemistry story. For example, one glaring absence in the primordial soup was any equivalent of the ubiquitous fats that are such an essential part of living cells today. Although these tarry substances contained no fats, some of their molecules had properties that resemble those of fat molecules. As we will see, such molecules certainly played a central role in life's origins.

Such a primordial oil slick would not have lasted for long, however. It would quickly have congealed into tar and then been broken down slowly by sunlight. Even this mundane process could have been an important component in the synthesis of organic material. Today, the degradation of asphalt on roads and parking lots by sunlight contributes a significant quantity of methane to the total amount of this now-dangerous gas building up in the Earth's atmosphere. A similar process on the early Earth may have been important in converting the primordial oily material into the methane needed to drive prebiotic reactions.

The oily goo is one promising source of organic materials. But there may be others. Some scientists have suggested that volcanoes, geysers, and hydrothermal vents in the deep ocean could have provided the answer, although this idea too has been besieged by controversy.

Ventists

In February 1977, a dramatic discovery was made by the research submarine *Alvin* as it dove in the deep ocean off the Galápagos Islands. The crew of the *Alvin* was familiar with the deep ocean world from many previous dives, but what they saw on this dive was utterly new.

The ocean floor below the reach of sunlight is normally covered by featureless, gray ooze, with only the occasional large animal or burrowing worm to relieve the monotony. On this history-making dive, the submarine cruised cautiously over just such an ooze-covered plain, toward a volcanically active region where hot, mineral-laden water was known to be spewing out of geothermal vents.

Such vents are common along the midocean ridges that girdle the world. At these ridges the crust is thin. This permits molten magma from below the crust to rise toward the surface, spreading the seafloor apart and powering continental drift. These vents are the marine equivalent of the similar vents on dry land that dot Yellowstone National Park and that are also found in many geothermally active parts of the world such as Iceland, New Zealand, and central Java.

On land, as the superheated water from these vents comes to the surface, it immediately turns to steam, laden with the hydrogen sulfide that gives the vents their typical rotten-egg smell. In the depths of the ocean, because of the high pressure, the hot water remains liquid even at temperatures in excess of 300°C.

FIGURE 4.3 Tube worms, a small crab and mussels (in the background, lower left) living around a deep-sea hydrothermal vent discharge. This region called the "Rose Garden" is on the Galápagos Rift side branch of the East Pacific Rise and is at a water depth of 2500 m. (Photo courtesy of Robert Hessler, Scripps Institution of Oceanography)

As the *Alvin* approached the vents, its lights revealed something very different from the usual featureless bottom. Complex and crowded communities of animals were growing in clusters around the dark, chimneylike openings of the vents. Red-and-white tube worms, sometimes 2.5 meters long, were everywhere. These rich worm gardens also contained a wide variety of organisms such as shrimps, crabs, and mussels that thrived in the warm water (Figure 4.3).

Alvin and later submersibles have brought up many samples of animals and bacteria from these communities. Some organisms show extreme adaptations to their strange environment. The tube worms living near the vents have lost both mouth and anus. We know that their ancestors once had these normally indispensable structures, because the worms' distant relatives in shallower water have functioning digestive systems. Because the vent worms are unable to feed, they must gain all their energy from the billions of bacteria that have colonized specialized structures found inside the worms' bodies.

Free-living bacteria can be seen multiplying in whitish clouds around the vents. These bacteria, and the bacteria living inside the tube worms, oxidize the hydrogen sulfide that spews from the vents. Because all the animals in these communities draw ultimately on the bacteria for nourishment, they are not directly dependent on solar energy.

The remarkable properties of these isolated ecosystems have spawned a group of researchers, known colloquially as ventists, who believe that hydrothermal vent systems must have been important in the origin of life. Arguments about the role that the vents might have played are intense and passionate, pitting the ventists against those who claim that the vents are more likely to destroy organic compounds than to create them.

Some ventists, notably Everett Shock at Washington University in Saint Louis and his co-workers, have developed elaborate theoretical schemes to show how key organic compounds could have been produced directly in the hot vent waters. They calculate that chemical reactions in the superheated water, starting with carbon dioxide and other simple molecules and catalyzed by various minerals, should lead to the generation of a wide variety of amino acids, aldehydes, and ketones.

Other ventists, such as Jack Corliss, now at the Central European University in Budapest, Hungary, and John Baross and Jody Deming of the University of Washington in Seattle, have offered even bolder ideas. They speculate that rapid cooling of the scorching hot vent waters as they come in contact with the surrounding cold, deep ocean water should assemble simple organic molecules into more complex ones. The net product of these reactions would have been nothing less than life itself.

A major problem with all these theories is the destructive power of the vents themselves. The water spewing from the vents actually reaches more than 350°C before it is diluted with surrounding seawater. This high temperature rapidly destroys organic compounds rather than synthesizing them.

At these temperatures, amino acids undergo total decomposition on a time scale of only a few minutes. DNA is broken apart into small fragments in seconds. Indeed, when samples of water from hydrothermal vents are analyzed, no amino acids are detectable, even though they were present in measurable concentrations in the seawater before it was transported through the hot vent systems.

It can be calculated that virtually *all* the water in the oceans must circulate through hydrothermal vents every 10 million years. Because vents were probably more plentiful on the early Earth than they are today, the cycling time was probably even shorter. As a result, organic compounds could not have built up over long periods in the oceans without being destroyed, although they could of course have accumulated locally on rocks and in ponds.

This problem of decomposition has not deterred the ventists, however. Recently, instead of advocating the synthesis of organic compounds directly in the scalding hot vent waters, they have suggested that it took place instead on the fringes, where vent waters have had a chance to cool.

This idea has been championed by Günter Wächtershäuser, an organic chemist turned patent lawyer in Munich, Germany. He has concentrated on the role of pyrite, a simple compound of iron and sulfur. This mineral precipitates in large quantities near the mouths of present-day vents and likely did so in the past as well. It forms black, crusty deposits very different in appearance from the glittering fool's gold crystals that most of us think of when we think of pyrites. It was this material that built up into the chimneylike structures that loomed in *Alvin*'s spotlights when the submersible first arrived at the vents.

These deposits, with their wide variety of chemically reactive surfaces, turn out to be much better catalysts for a variety of reactions than the regular surfaces of fool's gold. Wächtershäuser has succeeded in converting carbon monoxide into acetic acid under laboratory conditions that are supposed to mimic some of the conditions near vent discharges. He also finds that pyrite can catalyze the linking together of amino acids into short chains. In essence, he has discovered another way to produce a set of molecules like the proteinoids of Sidney Fox, but by using various finely divided pyrites as catalysts.

Wächtershäuser, like Freeman Dyson, is a metabolist, believing that metabolic pathways must have preceded the appearance of genes. He speculates that the regions around the vents would have acted as organic synthesis factories, producing molecules that could in turn help to catalyze key steps in a series of continuously operating reactions. According to Wächtershäuser, this series of reactions would have been a primitive type of metabolism, and according to his criterion, the system at this point would have been living. The reactions and compounds needed for accurate replication and the appearance of molecules carrying genetic information would have come later.

In collaboration with a chemist, Claudia Huber, Wächtershäuser has conducted some experiments bearing on his proposals. They are exciting experiments, because they explore the role of pyrite as an important inorganic catalyst. Unfortunately, they are bedeviled by the same problem that arises with other prebiotic experiments: They require very high concentrations of building blocks at the outset.

In addition, some of the steps that Wächtershäuser proposes in his syntheses seem not to work very well. Subsequent experiments carried out by a group of researchers (including Stanley Miller and one of the authors of this book, JLB) at the University of California at San Diego, using the conditions that Wächtershäuser suggested, failed to find detectable amounts of amino acids or other key compounds such as the nucleobases of DNA and RNA.

Theoretical calculations by Martin Schoonen and co-workers at the State University of New York at Stony Brook show that the concentrations of amino acids produced by a synthesis of this type would be so low that a liter of seawater would contain only a few million amino acid molecules. This may sound like a lot, but it isn't—10 million amino acid molecules in a liter of water is only one part in 5 quadrillion, a very dilute soup indeed.

Even supposing that the Wächtershäuser syntheses could have taken place under special conditions and produced an abundance of amino acids and other organic compounds, there is no reason that they would have had to take place near the hydrothermal vent itself. If the proposed reactions could have occurred in vent water cooled to, say, 100°C, why couldn't they also have occurred, albeit more slowly, in vent waters cooled to, say, 50° or 25°C? Certainly these temperatures would also have existed near the vents. And why wouldn't the same set of reactions have been just as likely to occur in some tidal pool on the rocky shore of an island, so long as pyrite was present? At what temperature is a Wächtershäuser-type reaction scheme no longer a "hydrothermal" process?

In spite of these problems, vents undoubtedly have interesting properties and may offer a route for the synthesis of some organic compounds. For example, Koichiro Matsuno and his co-workers in Nagaoka, Japan, have recently tried to imitate conditions at a hydrothermal vent in the laboratory, to see whether peptides could be synthesized. They passed solutions containing dissolved copper and glycine through a catalytic system that was repeatedly heated and cooled rapidly, and obtained a variety of peptides with up to six glycine molecules joined together.

They needed a lot of glycine to drive the reaction, and they heated the glycine solutions only briefly before cooling them again. This brief heating would not of course have been the case in those ancient hydrothermal deep-ocean systems. Karl Turkian at Yale University has estimated that the seawater in hydrothermal vent systems is heated to temperatures in excess of 350°C for more than thirty-five years. But Matsuno's conditions might have been found near periodic underwater geysers, the submarine equivalent of Old Faithful, where the water was repeatedly heated and cooled on a short timescale.

In sum, vents are intriguing, but the big problem with them is that they do terrible things to the primordial soup. As soon as prebiotic synthesis diminished or stopped, they would have begun to destroy the rich mixture of organic materials in the oceans of the early Earth. But perhaps there were ways that components of the primordial soup could have been protected from vent destruction. For example, some organic compounds were probably removed from the soup as they adhered to the surfaces of rocks and sand particles, which prevented them from being cycled again through the vents. Such surfaces were also promising places for some of the other steps leading to the origin of life.

Life on the Rocks

In 1948, the physicist J. D. Bernal wrote an essay in which he suggested that the catalytic synthesis of organic compounds on the surfaces of clays and other minerals—perhaps on some primitive beach or mudflat—may have played a role in the origin of life. Some scientists have gone much further and have even claimed that the first life on Earth was actually at least in part based on minerals. Life that consisted entirely of organic molecules would only have appeared later.

Graham Cairns-Smith of the University of Glasgow first proposed in the 1970s that the original genetic material could have been a mineral crystal, such as those found in clays. The surfaces of crystals, with their aligned rows of molecules, could have provided templates for organic materials to align and orient themselves. These surfaces would have been the first genes, later to be replaced by more advanced genetic material. Such a replacement, by analogy with the world of commerce, he termed a "genetic takeover."

Unfortunately for his idea, it quickly became apparent that minerals simply could not carry enough information to be the genetic material for even the most primitive life. This mineral-first hypothesis also faces another severe problem: If clays were once the basis for life, why aren't traces of them still inside us? Except perhaps in the hands of animators who use clay to shape singing raisins and other figures for cartoons, there is no sign of these materials in the biosphere today. Although our cells contain echoes of other minerals, notably pyrite, there is no trace of clay.

Cairns-Smith's notions about clay genes have now been largely rejected by the scientific community. Two important concepts that he championed still remain viable, however. One, which we will explore later, is his idea of a genetic takeover. Although the first genes were certainly not minerals, they need not have been DNA or even RNA. They might have been something simpler, later to be replaced by RNA and subsequently DNA.

The second of his durable concepts is that some minerals, even though they are not information-rich enough to be genes, are excellent catalysts. Stanley Miller's first apparatus consisted of bare glassware, a smooth and uniform surface that yielded very few opportunities for chemical catalysis. But the surfaces of minerals provide a much more complex environment.

Cairns-Smith has pointed out that there is an incredible world of mineral surfaces that molecules can perceive even if we cannot. One of these is a hydrated aluminosilicate clay known as montmorillonite, named after a town in France where large deposits of this material have been found. The clay, each gram of which contains millions of tissue-thin mineral layers, is widely used in industry as a catalyst for organic chemical reactions.

Many efforts have been made to test the importance of such minerals in either direct organic synthesis or in polymerization reactions. Because certain minerals have charged surfaces, they can bind organic molecules. Once bound to the mineral, the molecules are no longer simply free-floating in solution but have been brought together in specific orientations. This makes them more likely to interact with each other. Mineral surfaces have been suggested to act like very primitive enzymes, able to promote chemically difficult processes like the polymerization or linking together of amino acids or other simple subunits into a chain.

Using this mineral-binding approach, Leslie Orgel of the Salk Institute for Biological Studies in La Jolla, California, has succeeded in making nucleic acid polymers consisting of twenty or more subunits. Indeed, using this polymerization-on-the-rocks approach, Orgel has found that once the polymers reach a certain length, they seem to be able to grow indefinitely.

All this seems almost too good to be true. And it is. First, once the polymers get long enough, they tend to be attached so firmly to the crystal surface that it is almost impossible to get them off. Second, the subunits used in these experiments must be "activated" beforehand in order for them to polymerize. Orgel and others have used nasty-sounding compounds such as methylimidazolide and carbonyldiimidizole as the activation reagents. These reagents would certainly not have been present in the primordial soup. How, in their absence, might the subunits have become activated?

Other chemicals present in abundance on the early Earth may have played a role in the activation process. One class of these is a set of compounds known as thioesters, sulfur-containing molecules that, like pyrite, carry a whiff of their volcanic origins.

Thioesters are sulfur-containing derivatives of simple organic acids such as acetic acid. A plausible role for thioesters in prebiotic chemistry and polymerization reactions has been elaborated in detail by Nobel laureate Christian de Duve in his book *Blueprint for a Cell: The Nature and Origin of Life*.

In the thioester world that de Duve proposes, organic acids were converted into thioesters in the sulfide-rich regions near volcanoes. These thioesters could trigger the formation of amino acid multimers. De Duve makes a judicious choice of the word *multimer* rather than polymer, because "polymer" suggests a regular chain such as those presently found in proteins and nucleic acids. Multimer, on the other hand, simply means a molecule with many parts. Fox's proteinoids, with their tendency to branch and clump, are multimers rather than polymers.

Like Fox before him, de Duve thinks that some of these multimers could have played a role in the formation of life. But he goes a step further, suggesting a plausible mechanism. Some of these molecules would form insoluble aggregates. They would tend to bind to other compounds in the environment, and this synergistic relationship would act to protect the multimers from degradation. This protection would thus allow particular types of multimers to accumulate, beginning the process of chemical evolution.

The ideas of de Duve offer not only a potential way to make simple polymers on the primitive Earth, but also a means by which some of the polymers may have acquired a primitive kind of catalytic activity. They are not very different from Fox's ideas, but with the addition of some interesting catalytic chemistry.

But what about the activating compounds that would be needed to start this whole process, the thioesters? Could they have been made under plausible primitive-Earth conditions? Wächtershäuser and Huber have done some fascinating experiments that suggests they might have been. The reaction mixture that they used started with carbon monoxide and methyl mercaptan, another sulfur compound, which is so strong-smelling that gas companies add it in trace amounts to natural gas so that people will be sure to notice gas leaks. Methyl mercaptan has been reported to be present in volcanic gases, and although large quantities of it would certainly kill creatures like us, it is not poisonous to all living things. Indeed, some astonishing bacteria can actually use the horrid stuff as an energy source.

When Wächtershäuser and Huber allowed their mixture of methyl mercaptan and carbon monoxide to react at 100°C in the presence of nickel and iron sulfides, the thioester of acetic acid was produced. This thioester was then able to catalyze the formation of peptide bonds between amino acids. When amino acids were added to the reaction mixture, about 2 percent of them were linked together into short chains. As usual, there is a cautionary note: The concentration of amino acids used in this experiment was much higher than any that might have been expected in the early ocean. At lower concentrations, the yield of amino acid multimers would have been much less. Nonetheless, they would have been made.

Arthur Weber at the SETI Institute and the NASA Ames Research Center has developed an even better scheme, by making thioesters of amino acids themselves. When he reacted formaldehyde and ammonia under mild conditions in the presence of glycolaldehyde and a thiol such as methyl mercaptan, the amino acid thioesters that were produced promptly condensed into peptides. Weber also found that the peptide products in turn tended to catalyze amino acid thioester formation, so that the products acted as autocatalysts. Once set in motion, the mix continuously produced peptide products until one of the reactants was exhausted.

Thioester chemistry is becoming yet another rich lode of research ideas and experiments. It seems that sulfur, belching out of the vents and volcanoes of the early Earth and transformed into many highly reactive compounds, played a significant role in the origin of life.

Chemical catalysis on surfaces adds a new and promising dimension to origin-of-life studies. The exciting thing about this new catalytic chemistry is that it is possible to get away from the simple chemistry that must have taken place in the primordial soup. Assuming that there were at least some organic compounds around, and that not all of them had been gobbled up and destroyed by the hydrothermal vents or blasted out of existence by ultraviolet light, then they should have been able to come together on the surfaces of minerals in all kinds of interesting ways. As a result, the early Earth was probably abundantly supplied with polymers of a wide variety of types.

The trick now is to somehow separate these polymers, getting rid of the uninteresting ones and retaining those with important functions that might have aided in the formation of life. It is time to wrestle with this sorting-out problem.

SORTING OUT THE GEMISCH

From his brimstone bed at break of day
A walking the devil is gone,
To visit his snug little farm the earth
And see how his stock goes on.

                                    Robert Southey

SUPPOSE THAT YOU ARE ABLE TO GO BACK IN TIME TO ABOUT four billion years ago and walk along an island beach of the early Earth. You will find it a breathtaking experience in more ways than one. First of all, you must wear a space suit, for there is no oxygen in the atmosphere. And the suit will have to be Teflon coated, because nasty gases present in the thick air, such as hydrogen sulfide and the vapors of sulfuric and hydrochloric acids, would certainly kill you very quickly by eating through any protective suit made of ordinary materials.

Can you see your surroundings? Not very well—the atmosphere is so smoggy that little or no light can penetrate to the surface, even at high noon. But lightning flashes do give you brief, lurid glimpses of your immediate neighborhood. And you might be able to see the red glow of a nearby active volcano.

A vast ocean stretches away into the darkness and murk. During the brief lightning flashes, you can see that it is covered by lumps of oily material. The color of the water between the lumps is a muddy reddish brown, stained by large quantities of organic substances. Great waves crash on the shore, built up unimpeded to awesome size as they cross the vast stretches of ocean that girdle the planet. As they break, strong winds whip the sea up into a roiling mass of foam. Gusts of rain slash across the landscape.

Winds and tides together have produced the huge, sandy intertidal zone across which you are trudging. The sand is of many colors—black, red, yellow. Even through your spacesuit you can hear a great hissing noise as the water withdraws after each wave. As you look more closely, you see that the rocks and sand are encrusted with glistening layers of organic material of many different colors that has been thrown up by the wind and tides. The layers hardly have time to dry out before the next tide sweeps back up the beach.

This ocean water of four billion years ago is salty, perhaps even more than today's oceans, but it also contains quantities of toxic compounds such as hydrogen cyanide and formaldehyde. Like a single breath of the atmosphere, a small amount would kill you.

You are not enjoying your stroll. The world is a noisy, dangerous, and utterly alien place. You can walk no further because, not far down the beach, a river of lava runs down to the sea, hissing and boiling and throwing up clouds of steam.

Captain FitzRoy of Darwin's ship the *Beagle* described his landfall on the Galápagos Islands as a shore fit for pandemonium. But he could not have imagined the truly hellish landscape of the early Earth.

Enter Charles Darwin, Hesitantly

The picture is one of utter confusion at the molecular level as well. The chemistry that took place on the Earth during its early history was both frustratingly limited and very dirty. It is difficult in the extreme to imagine how anything like the remarkable degree of molecular order found in today's living organisms might have arisen. Whole classes of compounds important to life were missing or, at the most, present in infinitesimal amounts. And many other compounds that were common then, like some of the unusual amino acids that Miller detected in his experiments and that are found in the Murchison meteorite, are not found in living organisms today. How could the rare have become common, the common have become rare, and new and highly specific kinds of compounds have been synthesized, even before there was life?

In spite of all the prebiotic syntheses that have been carried out in the laboratory, there has been a frustrating shortage of some of the most important building blocks of life. Experiments of the Miller-Ury type have been able to produce nothing that looks like chlorophyll or any other similar compound that might have been able to trap the energy of sunlight. Fats as we know them are not present, although some molecules in the oily scum that is the chief product of these experiments do have fatlike properties and form films and bubbles. It is difficult to imagine life as we know it existing without the help of at least some of these compounds. Nevertheless, we must come up with scenarios for how life could have begun in their absence.

To begin with, compounds important in biology today may not have been as important for the first living organisms, and vice versa. When they first appeared, living organisms might have used some compounds that were abundant in the Earth's primordial soup but that are no longer found, or are found only rarely, today in biology. In fact, some compounds that play a major role in biochemistry today may have been present in the primordial soup at very low concentrations, not because they did not form, but because they were very unstable and decomposed soon after they were synthesized.

One example is ribose, the sugar component of RNA. In 1861, the Russian chemist Alexander Butlerov showed how various sugars, including ribose, can be synthesized from concentrated formaldehyde solutions containing mineral catalysts such as carbonate and alumina. However, ribose is very unstable and tends to break down over a matter of days when gently heated.

Another example is cytosine, an essential nucleobase of RNA and DNA. Although other nucleobases, including adenine, can persist for thousands of years in solution, cytosine is unstable and decomposes quickly. This problem has suggested to many chemists that the first genetic material did not contain cytosine.

Of course, this does not rule out the possibility that a few of these unstable molecules could have existed at any one time and that they might by chance have joined together and formed some sort of primitive genetic material. As little as one molecule that has the ability to make copies of itself can go a long way. Once even a single self-replicating molecule arises, it can quickly multiply to enormous numbers through geometric progression—one molecule gives rise to two, which give rise to four, and so on. Unless such a molecule is extremely unstable, its ability to make copies of itself rapidly should allow it to keep ahead of any breakdown resulting from its instability.

Getting to such a self-replicating molecule, however, is very difficult. And the first molecule that had talents in this direction would have needed plentiful building blocks of very specific types and copious quantities of chemical energy if it were to replicate itself successfully. Unfortunately, the primordial soup would have been a poor place for this. It would have been like a builder's yard that has only one or two examples of every imaginable type of brick, board, and plumbing fixture—a wonderful collection for browsing, but not much good for building a house.

The soup does show some differences from a builder's yard, however, for some of the molecular components of the soup have the remarkable capacity to assemble themselves in interesting ways. Oily molecules can come together to form globules, structures with an inside and an outside. Peptides and similar molecules can become concentrated within the globules. Acidic molecules can stick to alkaline ones. But these various kinds of molecules must be brought together and concentrated for this to happen.

At many stages during the origin of life, powerful sorting-out processes must have been at work. These sorting processes winnowed out the molecules most likely to lead to life from that plethora of builders-yard choices, and allowed them to interact. It is at this puzzling junction, four billion years ago on a still-lifeless Earth, that the shy, bearded figure of Darwin must have made his first appearance.

But how can Darwin have begun his work in a world made up only of collections of organic molecules, before the beginning of life? It is time now for us to examine critically Aleksandr Oparin's idea of chemical evolution and to try to tie it more firmly to Darwinian evolution.

Deconstructing Darwin

Early in the twentieth century, the philosopher Henri Bergson proposed that a mystical property distinguishes life from nonliving things, a property that he called *élan vital*. Going further, he suggested that this property somehow drives evolution. But his evolution was not Darwinian. Instead it was a directional process, built into the very nature of life, which led to greater and greater complexity. His élan vital was very like the "felt needs" that his countryman Lamarck had earlier imagined that animals possessed and that drove their evolution.

There is, of course, no evidence for such a life force, and these days scientists realize that all the phenomena Bergson attributed to the life force can be explained by Darwinian evolution. But there is something beguiling about the concept, and Aleksandr Oparin was not immune to it when he proposed his idea of chemical evolution.

Oparin claimed that nonliving molecules could somehow evolve, because the more persistent molecules would survive. But minerals persist for billions of years, and they do not evolve. Surely, nonliving molecules do not have an élan vital. Something else is needed. The obvious way out of this impasse, as Oparin realized, is to extend the influence of evolution back to a time before there were any organisms that we would define as living.

Surely, however, evolution in the absence of organisms must be an oxymoron! How could a prebiotic evolutionary world possibly work? It is here that Oparin was reduced to hand-waving. We think we can do a little better.

Obviously, the process of evolution had to begin somewhere, and like life itself, it must have started out very simply. We can begin by asking what the absolute minimum requirements must be for some kind of evolution to take place. We must deconstruct Darwinian evolution into its component parts and see which of them could have applied to the nonliving world.

Darwinian evolution is extremely good at sorting things out. At present, it sorts out genes, separating favorable genes from unfavorable ones. Because the organisms that carry the favorable genes are more successful reproducers, their genes are represented by more copies in the next generation. Less successful genes disappear. We always talk about evolution and genes together in the same breath, because it is the selection of one kind of gene over another that brings about evolutionary change.

It is difficult for us to imagine any sort of Darwinian evolution in the absence of genes. But if the sorting-out capability could somehow be decoupled from the reproductive capability, then at least some aspects of Darwinian evolution might apply to molecules other than genes.

Consider what we have learned about the world as it existed shortly before the appearance of life. Darwin's own hesitant suggestion that life might have begun in a warm little pond has become modified beyond recognition by the work of subsequent generations of astronomers, geologists, and chemists. In actuality, there were really rather few places anywhere on the surface of the storm-lashed, earthquake-wracked, volcano-studded early Earth that would have permitted the existence of such quiet little ponds. The environment was shaped instead by the tidal forces and giant cyclonic storms that swept almost unimpeded from one side of the planet to the other.

Nonetheless, chaotic though the early Earth would have appeared to our eyes, there was a surprising amount of predictability about the forces that shaped it. In particular, the tides, and probably even the storms, occurred on a regular schedule.

SORTING OUT THE GEMISCH

The result was that ocean currents driven by the Earth's rapid rotation roared back and forth past the scattered islands, many of which were capped by active volcanoes. Headlands jutting out from the coasts were interspersed with beaches and flat regions of sand and mud that were swept twice every day by substantial tides. In the tidal flats, runoff from the islands mixed with ocean waters and, combined with local evaporation, produced a scattering of fresh and salty lagoons that were probably spiked with different amounts of organic material. Then, a few hours or a few days later, tides and storms would sweep all this away and the whole cycle would begin again.

Because of that cyclic predictability, collections of molecules that could best resist being swept away by wave and tidal action would have accumulated. The tides and storms provided ample opportunity for these early accumulations of molecules to collect, be washed away, collect again, and become sorted out over and over.

Nearby, periodic geysers of mineral-rich and gas-laden water drained into pools and streams. As the water cooled and the minerals precipitated, a great variety of different environments were produced. You can see something like this today in the form of the mineral-laden pools that sparkle like chains of pastel-colored stepping-stones along the streams that drain the hot springs in Yellowstone National Park.

All this activity was not confined to the surface of the Earth. Tidal forces, and earthquakes resulting from the beginnings of tectonic movement, were introducing cracks into the masses of rocks that lay beneath the islands. This allowed water and dissolved organic materials to percolate down into the subsurface. There they met and mixed with hydrogen and methane, along with the more exotic gases such as the mercaptans, which were being released by heat and pressure from deeper rock formations. Beneath the sea as well, the vents spewed out fresh lava and huge quantities of mineral-laden water along with gases that had been added from the deep rocks.

In the deep-sea world, however, molecules could be washed away, never to return. And in the deep rocks far below the surface of the land, the percolation of water and gases was very slow. It was on the surface of the land, short-lived though landmasses were in those days, that the real action took place.

In the unstable and rapidly changing environments of the intertidal zones and geysers, collections of organic molecules would have periodically accumulated and interacted with each other. There they awaited the firm hand of Darwin, who would decide which of them would "live" or "die." Darwin provided the selective process that sorted out the more successful from the less successful of this diverse collection of molecules.

A successful organism is one that has many offspring. But how, in a prebiotic world, can we define success? Why should one collection of molecules have been more successful than others? Is success simply a matter of commonness? By that measure, the common mineral quartz is more successful than the rare mineral diamond! In the prebiotic world, the word *success* has no obvious meaning.

Nonetheless, throughout the planet, many factors must have determined which molecules would succeed. As the tide ebbed and flowed, some molecules and collections of molecules tended to cling to sand and mud grains of various compositions. Some of these molecules were subsequently washed away, particularly during storms, when the scouring effect of the tides was supplemented by strong wave action. Other molecules, because they were unstable, simply decomposed.

Regardless of whether the molecules were washed off or decayed, their places would soon have been taken by other molecules that could cling more tightly. These molecules would have come from the immense supply of small molecules, of every imaginable type, that were floating in the primordial soup or that were being generated by catalytic processes on the surfaces of nearby rocks and minerals.

The result would have been a very primitive equivalent of birth and death. One deciding factor would have been the ability to cling to a particular type of rock in the face of those continual hosings from tides and waves and the occasional bursts of steam from nearby volcanic vents. Another would be how quickly the molecules could percolate through sand of various compositions.

Note again that this can only have been the beginning of Darwinian evolution. There was as yet no replication, and as a result there was none of the selectable genetic variation that would have been produced by the replication process. But there were at least the molecular equivalents of birth and death, which represented a start toward a more full-fledged Darwinian world. This was a world of "Darwin Lite."

As the sorting-out process continued over long periods, the resulting collections of molecules would inevitably have become more complex, even before there were any mechanisms for replication. If, as de Duve has suggested, chemical reactions took place between some of these molecules that allowed them to cling more tightly, then these new and more complex molecules would tend to replace the older and less

SORTING OUT THE GEMISCH

firmly attached ones. And as these groups of molecules grew in size, those able to elbow the neighboring molecular clusters from their perches would also have had an advantage.

Because the tides and storms came and went so dramatically and yet so predictably over short periods, it would not have been long before this rhythm would have sorted out organic molecules into stratified bands that occupied various parts of the tidal zones. Even the sand and mud particles themselves would have been sorted out through tidal action, the grains becoming stratified according to their sizes and chemical properties. Different collections of molecules would have clung to different types of particles. During your stroll on the ancient beach, you saw the result of these processes as layers of different-colored material that coated the sand and the nearby rocks.

To the untutored eye, of course, these collections of organic material would have simply looked like scum or slime. But something very important would have been happening. Even though each layer was still chemically very diverse, within each layer, only certain kinds of molecular groupings would have been selected for, out of an infinite universe of possibilities.

The Buildup of Chemical Complexity

We do not know, because no equivalent laboratory experiment has yet been tried, just how complicated such collections of molecules could have become over thousands or tens of thousands of tidal cycles, and what kinds of properties they might have acquired. The collections of molecules in a given strip of the intertidal zone would at least have shared certain basic physical characteristics—at a very minimum, the ability to cling to a particular kind of particle under certain conditions of wetting and drying without being washed away or being broken down by sudden blasts of ultraviolet light from the sky. And the most successful of them might also have been able to repel any molecular invaders of their little realms, perhaps by presenting a slippery, slimy coat to the outside world or perhaps by possessing catalytic properties that could actually destroy the invaders. This may seem unlikely, but remember that even a single sand grain would have had enough room on its surface for thousands of such evolutionary "experiments."

These experiments would not have been confined to the rocks and beaches. As we saw earlier, the streams and pools that drain the water from today's periodic geysers provide a striking set of graduated environments that sort out the organisms that live in them. The geysers of the ancient world must have had these properties as well.

Some heat-resistant bacteria can grow very close to the vents of today's geysers, because they can withstand the periodic dousings with hot, mineral-laden water as it condenses from the superheated steam. Just as such extremely tough bacteria are selected for at the present time, on the early Earth the clusters of molecules that could "survive" such extreme conditions would have been selected for as well. Some would have been periodically scoured from the rocks by particularly violent eruptions of the geysers, just as rocks and sand on the beaches would have been scoured by storms. But the survivors of these scourings would soon have been joined by molecules with similar properties that were able to "colonize" the areas that had been opened up.

Other chemical processes were at work. Most sands of today are made up of crushed silicates or crushed fragments of calcium carbonate from coral and protozoan skeletons. The sands of the early Earth were very different in their composition and included many minerals that would have been rapidly oxidized in today's oxygen-rich atmosphere. Some of these are the iron sulfides, which could have served as powerful catalysts for many different chemical reactions. It is unlikely to be a coincidence that compounds of iron and sulfur lie at the heart of many of the most primitive and fundamental biochemical processes taking place in the living cells of today (see Chapter 7).

If this proto-Darwinian evolution was to have taken place, there must have been plentiful sources of organic molecules nearby. These would have provided the rich streams of new organic chemicals that washed past and over the slimy layers on the rocks and in the sand and mud, adding to them, interacting with them, sometimes displacing parts of them. One such source would have been the quieter tidal flats that were out of the reach of waves and that were flooded and drained twice a day. Evaporation would have concentrated the organic molecules, and different molecules would have accumulated in different parts of the flats. Chemicals that were quite diluted in the early ocean would have become concentrated in these flats.

Stanley Miller and his co-workers have recently mimicked tidal pools in the laboratory to show how some compounds that chemists have had difficulty synthesizing under simulated prebiotic conditions could have been made easily on the primitive Earth. When they evaporated a solution of urea and cyanacetaldehyde (which is produced in spark discharge experiments), large amounts of the nucleobase cytosine were produced. In control experiments that used more dilute solutions, only trace amounts of cytosine were made.

In an even more important set of experiments, Miller and his student Jim Cleaves have shown that fairly elaborate sequences of reactions leading to chemicals that are essential for life can occur spontaneously, even without the aid of protein enzymes. Remarkably, these reaction sequences often follow the same path taken in living cells today. Although the reactions need high concentrations of their components, these might have easily built up as tidal pools evaporated and dried out.

We do not yet know how much such processes enriched the primordial soup. All this new work suggests, however, that the accreting layers of molecules on the surfaces of sand grains and rocks had a far larger "cafeteria" of new molecules to choose from than even the most enthusiastic prebiotic chemists have yet been able to make.

The concentration of compounds in drying tidal pools could have had another important effect. In the polymerization-on-the-rocks experiments of Leslie Orgel, as the polymers became longer, they tended to be more and more tightly bound to the mineral surface. Water itself could not wash these long polymers off the surfaces, but a sufficiently strong salt solution could. On the early Earth, polymers made on mineral surfaces in a tide pool or shallow lagoon could have been released back into solution as the water evaporated and concentrated the salt in solution. This salt-induced release must have played a critical role in allowing a random mix of polymers that was generated in one environment to interact with a different random mix made in another environment.

The reaction between amino acids and sugars is another process that may have played a role in the development of chemical complexity. When a solution of amino acids and sugars is heated, the result is a brownish polymer. This reaction is the same as the one that causes your bread to turn brown when you make toast. It also takes place in people with diabetes and is one reason why their eyes tend to be so prone to the development of brown-pigmented cataracts.

In those early tidal flats, whenever amino acids and sugars were concentrated by evaporation, these brownish polymers would have formed readily. The polymeric products of this reaction would have roughly resembled the proteins on the surface of today's living cells, which tend to be covered with chains of sugar molecules. The accumulating sugars could have provided some sort of protective function—perhaps by producing a slimy impervious layer that would have repelled water as tides and waves periodically swept by.

It is striking that the most heavily sugar-coated protein molecules of today tend to be found on the outside of living cells. The various combinations of these sugars trigger most of our immune responses. Could the first prebiotic layers of molecules have become armored with sugars that protected them from the environment? And could this have led, through billions of years of later evolution, to the sugar-coated cells of today, and to our own highly sophisticated immune systems?

An Ancient Laboratory on the Beach

In the laboratories of today's chemists, glass columns filled with finely divided minerals are often used as sensitive and selective filters for separating out chemicals that might be quite similar to each other. Amino acids with similar chemical properties can easily be separated on such columns, and single-stranded DNA can be separated from double-stranded DNA. The process is called chromatography, because it was first used to separate mixtures of colored dyes and other colored compounds (Plate 8).

Chemists of today can also separate large protein polymers by a variant of the chromatographic technique called gel filtration. Porous gels, some made up of organic materials, can be used to separate large from small molecules. The smaller molecules tend to take a longer and more devious route through the pores in the gel fragments, so that the large molecules emerge from the bottom of the column first, followed in order by the smaller ones. These and innumerable other variants of the same basic process of chromatography have provided chemists with powerful and sensitive tools for all kinds of separations.

Chromatography could have happened in the prebiotic world as well. As layers of organic molecules accumulated, some of these layers could have acted as gels that performed chromatographic separations, retaining some types of molecules and releasing others. The number of possible chromatographic processes would have grown as the mix of organic molecules adhering to mineral surfaces grew in size and thickness. We might think of the slimy layers accreting on the tidal flats and beaches on the early Earth as vast geochromatography experiments, repeated twice a day for centuries or millennia.

We have strong evidence that geochromatography has had an important role in shaping the Earth that we know. For example, it is one way by which nature has sorted out the complex mixture of organic compounds in sediments during the formation of petroleum deposits.

Water that percolated through the sediments preferentially carried the soluble compounds away and left behind the less soluble compounds that make up petroleum.

A graduate student at the Scripps Institution of Oceanography, Michael Wing, has demonstrated the power of geochromatographic separation using materials that might have been found on the early Earth. He filled a glass column with beach sand and layered the top with a mixture of three slightly different types of a class of compounds called polycyclic aromatic hydrocarbons (PAHs). These molecules are components of the oily material found in the Murchison meteorite, and he chose three that are slightly water-soluble. As seawater was passed through the column, each of the three PAHs emerged from the bottom of the column at different times. This simple geochromatography experiment resulted in the almost complete separation of these chemically very similar compounds.

This kind of chromatographic resolution has produced highly purified PAHs under natural conditions. The mineral Pendletonite, found in deposits in California's San Benito County, has been found to contain inclusions of 99 percent pure coronene, a seven-ring PAH. The coronene found in this natural deposit is actually purer than the same compound available commercially from chemical companies. The naturally occurring coronene was produced by the geochromatographic resolution of a mixture that originally contained many different PAHs and other hydrocarbons.

The result of all these processes would have been layers of organic molecules that would have been constantly growing and changing, increasing in thickness and complexity. The layers would have consisted of a thicket of proteinlike molecules able to trap a variety of smaller molecules along with globules of tarry PAHs, all covered with a slimy layer of modified sugars. Although the layers were still a mixture of many different types of molecules, the mixture had acquired a definite structure that had been shaped by the "birth" and "death" process of proto-Darwinism. And they would be ready for the next step.

All of the various sorting-out processes undoubtedly played a role in the natural selection of molecules and produced an increasing level of molecular complexity and function. But even so, this prebiotic mixture of organic compounds would have not resembled those that we see in living organisms today. Further fine-tuning of the organic mixture must have taken place, either during the prebiotic period preceding the origin of life or during the early evolution of life itself.

Winners and Losers in the Molecular World

Stephen Kent and his colleagues at the Scripps Research Institute in La Jolla, California, have recently carried out an amazing chemical tour de force. They succeeded in making an enzyme, ninety-nine amino acids long, entirely from D-amino acids.

You will recall that in present-day organisms, only left-handed amino acids (L–amino acids) are used by living cells to construct proteins. The enzyme that Kent's group copied is one manufactured by cells infected with the AIDS virus. Its function is to cleave other proteins at specific points. Kent and his colleagues found that their mirror-image copy of this enzyme was perfectly capable of cleaving a peptide in just the same way as the normal enzyme. The big difference was that for their mirror-image enzyme to work, the peptide that it was to cleave also had to made of D–amino acids.

Their new enzyme was therefore only able to function in a Through-the-Looking-Glass world in which everything was reversed. Given this constraint, however, the enzyme functioned quite normally. This remarkable experiment shows that there is no apparent biochemical reason why L–amino acids would have been selected over D–amino acids when life began.

The decision point that determined which type of amino acid would be used by life must have happened very early. And many other kinds of decisions had to be made at the same time. Only twenty different amino acids are coded for by the DNA of modern organisms, even though a much larger variety of amino acids were present on the early Earth. The nucleobases adenine, guanine, cytosine, thymine, and uracil are the only ones used in the nucleic acids of modern organisms, even though many others could have been present in the Earth's prebiotic soup. How did this select set of organic compounds come to be chosen?

There are no definitive answers to these questions. Some researchers argue that there had to be a prebiotic selection process that favored the molecules that make up all life on the Earth today. Others argue that it was life itself that culled out the molecules currently used in biochemistry out of the vast array originally present, and that some choices were simply a matter of chance—a frozen accident, as Francis Crick has described it.

The various nucleobases found in RNA and DNA were probably selected for because they readily form pairs with each other. In the Watson-Crick base-pairing that is central to the structure of DNA, adenine pairs with thymine and guanine pairs with cytosine. In RNA, uracil substitutes for thymine, and it too can readily bond with adenine. This base-pairing is what holds the DNA together in a double helix.

However, the hydrogen bonds that form the pairs must not be so strong that they cannot be broken or so weak that they fall apart too easily. If these pairings cannot be readily broken after they have formed, the machinery of the cell will soon grind to a frozen halt. Many nucleobases other than the ones used in biology have been tested, and it has been found that they either form base pairs only with molecules like themselves or form interactions with other bases that are either too strong or too weak. As a result, these bases were not used or were discarded during the early evolution of the genetic material.

It is also fairly easy to explain why life today is based only on the twenty amino acids commonly found in proteins. These amino acids have structures that lend themselves to formation of peptides in regular chains, with side groups that exhibit a diversity of different functions. Most of the many other amino acids that were made in the Miller-Urey experiment, as well as those found in meteorites such as Murchison, tend to have chemically rather similar and uninteresting side groups. Worse, they do not join together to form regular peptide chains. Chains formed from these amino acids would have been branched, jagged, and clumpy structures. There must have been a great deal of selection among these early proteinlike molecules for those that had sufficient regularity to enable them to perform useful chemical tasks.

Handedness is a tougher problem. It is possible that the earliest organisms contained both all-D and all-L enzymes. But this means that they would have needed separate sets of genetic material to make the two types of enzymes. Any such redundancy would soon have been eliminated. Amino acids with one handedness probably became predominant very early on, and by chance it was the one based on L–amino acids.

Some scientists claim, however, that the L–amino acid handedness was preordained and that it was in fact necessary for the origin of life to take place. Various mechanisms have been proposed by which a racemic mixture of amino acids could have been changed into one in which only L-amino acids dominated. Up until recently there has been no direct proof that this could have happened.

Then, in 1996, John Cronin and Sandra Pizzarello at Arizona State University found that some unusual nonprotein amino acids in the Murchison meteorite show a small excess of the L-isomer (on the order of a few percent). How this excess was generated is uncertain, although Cronin and Pizzarello favor an idea first suggested by William Bonner at Stanford University. Bonner suggests that circularly polarized light associated with the radiation of neutron stars could have caused asymmetric organic synthesis.

Could such small L–amino acid excesses be the cause of the exclusive use of L–amino acids by life on our planet? For this to be the case, amino acids must have been delivered to the Earth from space without extensive decomposition. Then the small excesses must have been amplified by some unknown process, resulting in a mixture in which the L-isomers were the dominant ones present on the Earth. Finally, the predominance of L–amino acids would need to be somehow transferred from the nonprotein amino acids to those found in proteins today.

All this seems more than a little unlikely and farfetched. But the idea can eventually be tested. It predicts that if life exists elsewhere in our Solar System, it should also be based on L–amino acids. If we find life on Mars or Jupiter's moon Europa that is based on proteins and enzymes made up of only D–amino acids, like the enzyme that Kent and his colleagues manufactured, this would eliminate the argument that amino acid handedness in terrestrial life was preordained and was the result of some sort of abiotic asymmetric process.

The reason that D-sugars were selected by our remote ancestors is essentially unknown. Some researchers have argued that there is a relationship between the handedness of sugars and that of amino acids. Thus, the exclusive presence of L–amino acids in proteins would somehow dictate that D-sugars were used in nucleic acids or vice versa. No convincing experiments have shown this relationship. We do know one thing, however. The use of D-sugars could not have been the result of seeding the Earth with material from space, because there is no sign of sugars in meteorites. The most likely reason for the current use of D-sugars is that it was indeed a frozen accident.

Advancing Toward the Beginning

Scientists are now realizing that the environments in which early life might have appeared were far more diverse than could have been imagined just a few years ago. And, perhaps most importantly, these various processes and environments provided a means by which the sorting out that is required by the earliest phase of Darwinian evolution could have taken place. Once this began, some of the elements of the minimalist definition of life that we set out at the beginning of this book would have been achieved.

Not all of the conditions were present, of course. There was a simulacrum of birth, in the form of new molecules being generated all the time by prebiotic syntheses. And there was a kind of death, as molecules and clusters of molecules were lost from a particular environmental niche and were replaced by other, more successful ones. The most successful occupiers of the niche would have "won," but only in a very narrow sense. We must again caution that full-fledged Darwinian evolution was not yet acting on these molecules, for there were no self-replicating systems with the ability to pass information about their own structures from one generation to the next.

It should be possible to investigate this almost-Darwinian world in the laboratory. The machinery needed to combine a Miller-Urey type of experiment with mechanical and chromatographic separations of the resulting compounds is not beyond the capabilities of clever glassblowers. Rather than a simple flask representing the ocean, one might have a tube filled with a variety of different, finely divided minerals of the kinds probably found on the early Earth. Water could then be forced up and down the tube in a cycle of a few minutes or hours in length, to imitate the tides. How would the organic molecules being generated in the Miller-Urey half of the experiment sort out in such a system? Would there be unexpected and complex interactions? A world of new observations awaits the clever and patient experimenter.

The scenario we have just discussed is a plausible one, and it is experimentally testable in many ways. But it only provides a very partial solution to the problem of the origin of life. Molecules may have been sorted out to different places in that turbulent early environment and may have accumulated in a variety of different combinations. A high degree of chemical complexity may have developed during the sorting-out process. But then what happened to these chemicals? Selection for the ability to cling to rocks is simply not enough to explain the origin of life. What separates natural abiotic chemistry from biochemistry?

As Darwin sweltered on that multicolored beach, with all kinds of interesting chemistry going on around him, he still had a lot of work to do.

THE FIRST PROTOBIONTS

The problem of describing the origin of life is curiously like its actual origin. In each case it seems likely that a number of attempts were made, and almost all of them did not work. Natural selection eliminated them.

J.B.S. Haldane, New Biology 16 (1954): 26

It is evident that once a self-replicating, mutating molecular aggregate arose, Darwinian natural selection became possible and the origin of life can be dated from this event. Unfortunately, it is this event about which we know the least.

Carl Sagan, Radiation Research 15 (1961): 177

PHYSICIST AND MATHEMATICIAN J. D. BERNAL, WHO SPENT most of his working life at Cambridge University, was never short of ideas; he threw them out in all directions like a firework. Some were brilliant—he was one of the founders of X-ray crystallography, which could be used to probe the inner structures of crystals and which led eventually to an understanding of the structures of proteins and DNA. Some, alas, were distressingly wrong. For years, as a good Marxist, Bernal embraced Trofim Lysenko's grotesque ideas about genetics. And he insisted to Francis Crick that there was no point in looking at crystals of DNA, which were likely to be so complicated that they would be impossible to understand.

Bernal was, however, one of the first scientists after Oparin to address the question of the origin of genetic material. In 1948, when he first wrote about the problem, and in 1951, when he published a tiny book, *The Physical Basis of Life,* he assumed that the genetic material must be proteins, and that the first appearance of proteins must have been the point of origin of life.

A mere two years later, after the discovery of the structure of DNA, Bernal's idea fell by the wayside. More recently, it has been revived, with a number of clever additions. But the question of whether the first genetic material was protein or something else, and the furious arguments about the nature of the first genes and about when and how they first appeared, are still at the very center of the origin-of-life question a half century later.

Recall the dangerous stroll that you took along that primeval beach in the last chapter. Suppose you were able to collect and take back to a modern laboratory some samples of the layers of slime that you found on the rocks. Even with the best analytic techniques, you would have a hard job deciding whether the slime contained any genes. You would recognize the presence of polymers and might suspect that some of them could be self-replicating, but if these polymers happen not to resemble either protein or DNA, they would prove frustratingly hard to analyze.

As we have seen, prebiotic conditions tend to produce random collections of multimers in bewildering variety, rather than a large number of nearly identical copies of a few types. Thus, you might get a strong hint that something interesting was going on in those slimy globs if you find many polymeric molecules with similar structures. If you make such a discovery, this would suggest that some kind of replication had been going on in the material. But the mechanism by which the replication was happening, its chemical basis, and when it began, would probably continue to elude you for some time.

Denied access to those globs, scientists have done their best to construct scenarios for the appearance of the first genes. None are completely convincing, though we can borrow bits from one or another to make a moderately coherent story. The major argument centers on just how chemically elaborate the prebiotic world must have been before some kind of genetic material could appear—the old argument about genes-first versus metabolism-first that we met in the introduction.

Let us begin with one idea that is a particular favorite of molecular biologists, the "naked gene on the beach." The model is simplicity itself. The first gene was able to arise in the primordial soup, or more probably as a result of catalysis on the surface of some mineral particles. Right from the beginning, it was capable of making copies of itself from the building blocks supplied by the primordial soup.

As this ingenious gene evolved further, it began to catalyze the synthesis of peptides and other compounds that aided in its replication. The genetic material grew in size and complexity, coding for more and more compounds, with which it surrounded itself.

Soon, the primordial soup became exhausted of its few essential building blocks. Luckily, by this time the naked gene had managed to clothe itself in some kind of protective membrane and was able to extract energy from the environment and make the building blocks itself. Such a membrane bounded structure was no longer dependent on the compounds in the primordial soup, but could make energy-rich compounds on its own. This permitted the gene to make copies of itself with greater ease. After that, of course, it was plain sailing. Darwinian evolution could then take control of the process.

The original self-replicating molecule was probably not DNA or RNA, because these molecules are far too complex. But it was not long before the early genetic material would have been replaced by these more efficient polymers, which evolved the ability to code the fitting together of a more and more complicated collection of molecules.

This is the standard genes-first view of the origin of life. It requires that the first genes were able to make copies of themselves. But self-replication is something that the genes of living cells cannot do at the present time on their own. Genes today need very elaborate molecules, such as protein catalysts (enzymes) and RNA, in order to replicate. They cannot do it all by themselves now, so why must we suppose that they ever could? Remembering this caveat, let us explore the genes-first world and see where it leads us.

The RNA World

In the early 1980s, Thomas Cech at the University of Colorado and Stanley Altman at Yale University independently made a discovery that forever changed our view of modern biochemistry. They found that some RNA molecules were able to catalyze chemical reactions. Altman found an enzyme in E. coli, made of RNA and protein, that still carried out its function even after all the protein had been stripped from it. Cech found that RNA molecules present in the protozoan Tetrahymena could tear themselves free from a longer piece of RNA by bending back on themselves and catalyzing breaks at the point where they joined on to the main RNA strand. It is as if a page in a book had developed the remarkable ability to detach itself from the book and float away on the wind.

Such catalytic capabilities were previously considered the exclusive domain of protein enzymes. Indeed, the most convincing part of the work of Cech and Altman was the researchers' ability to purify the RNA molecules and utterly rid them of any proteins that might have been responsible for the observed catalytic activity. They were jointly awarded the Nobel prize in chemistry in 1989 for this sensational discovery.

THE FIRST PROTOBIONTS

Cech named these catalytic RNA molecules *ribozymes*. These molecules were clearly different from enzymes, because they could carry genetic information and act as catalysts simultaneously. He suggested that such molecules might, in principle, be able to catalyze their own replication. The implications for the origin of life were immense. Here at last were candidates for that naked gene on the beach.

The idea that RNA-based replication and catalysis might have originated before protein synthesis became a major biochemical pathway had first been suggested independently in the late 1960s by Carl Woese of the University of Illinois and by Francis Crick and Leslie Orgel, both of whom are now at the Salk Institute. But they had all assumed that the ability of RNA to carry out chemical reactions has since become lost in the course of evolution. It now appeared, from the work of Cech and Altman, that at least some RNA catalytic activity is still preserved in modern organisms.

Soon after the discovery of catalytic RNA, Walter Gilbert of Harvard suggested that there had once been an RNA world, a period in the history of the Earth during which life consisted of nothing more than self-replicating RNA molecules. In that world, as Gilbert imagined it, the RNA molecules began to assemble themselves from building blocks in the primordial soup. Soon they acquired the ability to synthesize proteins. Because proteins were more efficient and versatile catalysts, they rapidly took over most of the catalytic functions of the RNA, while the RNA retained its replicative ability. Finally, DNA appeared and took over the ultimate information-storing capabilities of the evolving cell.

RNA still gives hints of its central role in the world of the past. It is an essential part of the machinery of protein synthesis. And some viruses use only RNA as their genetic material and do not bother with DNA at all.

The idea of an RNA world is now widely viewed as a breakthrough concept in the long debate over how life on Earth could have evolved. But it is very far from being demonstrated. Ribozymes are all very well, but can they do all the things that would have been necessary in an RNA world? After all, the ribozyme reactions that Cech and Altman had first discovered had nothing to do with self-replication, but only with the ability to break an RNA chain. What other kinds of reactions could ribozymes catalyze, and could they really be persuaded to self-replicate?

Some answers to this question have been obtained by researchers using a laboratory approach known as test-tube evolution. The first such evolutionary experiments using RNA had actually been carried out by Sol Spiegelman of Columbia University, two decades before the discovery of ribozymes. Spiegelman set up a simple laboratory experiment in which pieces of RNA from a tiny virus called Qβ were mixed with a protein enzyme—made by the virus—that catalyzed the replication of its RNA. The enzyme quickly made many copies of the RNA. Then some of the copies were transferred to a new test tube, where a fresh supply of enzyme awaited them. This was of course an extremely simple system—no virus was being produced, only the RNA of the virus.

Because he always selected for the RNA molecules that replicated most quickly and therefore appeared first in the test tube, Spiegelman was eventually able to evolve tiny RNA molecules, only about two hundred bases long, that could still become attached to the enzyme and be replicated, but that had lost virtually all their other properties. These molecules, because they were so short, could replicate very rapidly indeed. They became known as "Spiegelman monsters," because they took over the experiment like alien creatures in a science fiction story.

This remarkably rapid evolutionary process, accomplished as the result of a few dozen transfers from one test tube to the next, produced an RNA molecule that could do nothing except attach itself to the enzyme and be replicated. Although the ancestor of this pitiful remnant molecule had once carried all the information needed to make a functioning virus, all this extraneous information had been lost because of the merciless selection for sheer speed that the molecule had gone through. At the end of this process, it could only do one thing—replicate at high speed. And of course it could only do this with the aid of the highly evolved replication enzyme, which you will remember is made of protein and not RNA.

Spiegelman's experiment pointed the way toward another and even more remarkable set of experiments conducted by Manfred Eigen at the University of Göttingen, starting in 1993. Eigen is a startling polymath. Quiet and rather shy in demeanor, he has done path-breaking experiments in organic chemistry by finding ways to follow rapid reactions in great detail. He won the Nobel prize in chemistry for this work in 1967. Also a talented amateur pianist, Eigen has played with a number of orchestras. And he has had profound insights into the state of the prebiotic world, particularly in the shape of a theoretical demonstration that the first genes had to be very short and simple because otherwise, they would undergo too many random mutations and be destroyed.

THE FIRST PROTOBIONTS

He and his collaborators began to wonder what would happen if they put the Qβ enzyme in a test tube that contained only the building blocks of RNA but no RNA molecules. The obvious answer, of course, was that nothing would happen. But remarkably, when the enzyme was left at loose ends in this fashion, it began to construct short stretches of RNA that had no relation to each other or to the original Qβ RNA. Then it began to copy these stretches. Eventually, pieces of RNA about a hundred bases long appeared that could be replicated quite readily in this system.

This remarkable result is not, of course, the creation of life de novo in the test tube. The protein enzyme that Eigen used is a highly sophisticated construct, the result of billions of years of evolution. But it shows again the power of Darwin, the ability of selection to sort out molecules and—in this case—to produce molecules that are readily replicated by the enzyme.

The discovery of ribozymes has made possible other test-tube-evolution experiments. This new generation of experiments has the goal not of simply producing a molecule that had lost capabilities, as Spiegelman's early experiments had done, but of producing one that had gained new abilities. Researchers such as Thomas Cech, Gerald Joyce of the Scripps Research Foundation (Figure 6.1), Jack Szostak of Harvard, and a host of others have managed to evolve new ribozymes that have a wide repertoire of catalytic functions.

Their first step is to generate a large set of random RNA sequences, by using a machine designed to synthesize such molecules. This immense array of RNA molecules, like the nonsense typed on a myriad of typewriters by an infinitude of monkeys, conceals the occasional molecule with interesting properties. The trick is to use selection to sieve these few molecules out of this huge collection of (at least to the experimenter) useless molecules. The experimenter employs a repetitive selective process, in which only the molecules able to do the task that he or she has set are permitted to replicate and be amplified. The random mixture can thus be "evolved" to the point at which the catalytic molecules dominate.

Various ribozymes have been evolved. Some can catalyze the formation of peptide bonds between amino acids; others can use metal cofactors that are different from those used by the natural ribozymes. Still others can actually link themselves to amino acids.

What is most exciting is that some of these selected ribozymes can make copies of small parts of themselves. It is as if a book had acquired the ability to make a copy of its last few pages.

FIGURE 6.1 Gerald Joyce of the Scripps Research Institute with a model of a ribozyme. (James A. Sugar, Mill Valley, California)

It is this last function, the ability to self-replicate, that Joyce and Orgel call "the molecular biologists' dream." If a truly self-replicating molecule could be produced in the laboratory, a huge gap in our understanding of the origin of life would be closed. Joyce is sanguine about this possibility, because he is most impressed by the ability of evolution in the test tube to generate ribozymes with new functions.

Remarkable as the results from test-tube-evolution experiments are, these evolved ribozymes are still a very long way from the full realization of the molecular biologists' dream. Even the cleverest ribozyme yet produced can only copy short stretches of itself. It is very unlikely, we suspect, that a molecule can be selected for that could polymerize a copy of itself along its entire length without some kind of help.

However, we must remember that the discovery of ribozymes is less than two decades old and that the field of RNA test-tube evolution in the laboratory is still in its infancy. The number of researchers carrying out RNA test-tube evolution is growing rapidly, and it is hard to predict what new reactions and properties newly evolved RNA, and even DNA, molecules may possess in the near future. There is a great deal of support for this type of research. Even though much of it is directed at

THE FIRST PROTOBIONTS

potential biomedical applications, the spin-off implications for the question of the origin of life are vast. It is not surprising that, in 1994 an optimistic Orgel predicted the imminent demonstration of self-catalyzed RNA replication.

Perhaps we will see, within the next decade or so, whether Orgel's prophecy comes true. Let us assume for a moment that Orgel is correct and that sometime in the near future a researcher will tease out, from the large array of random RNA sequences lurking in a test tube, the one that has the ability to catalyze its own replication from simple components of the type found in the primordial soup.

At this point, many researchers would argue that life has been created in the laboratory. But would this be a reenactment of the origin of life as it might have taken place on the early Earth? Certainly not! A much larger problem will remain: Even if researchers eventually do create such an astonishing molecule in the laboratory, this is no guarantee that a similar molecule would ever have been synthesized in the primordial soup or on rock surfaces early in the history of our planet.

Our hunch is that the molecular biologists' dream of a simple, entirely self-replicating molecule emerging on its own from the primordial soup is an unlikely mechanism for the origin of life. Whatever the mechanism by which the first replication was accomplished in the prebiotic world, it must surely have been a cooperative enterprise involving more than one type of molecule, accompanied by substantial flows of energy from one molecule to another. After all, this is the way that the genetic material replicates itself at the present time, with DNA, RNA, and proteins all cooperating with each other. There almost certainly never was a self-sufficient, naked gene on a beach.

There is another problem. Any such self-replicating molecule is likely to be long and complicated if it is to carry out completely self-catalyzed replication. And, as Eigen had shown, such a large molecule would have been very unlikely to appear in the prebiotic world, because mutations would quickly destroy it.

Was There a Pre-RNA World?

DNA is tough, but RNA is not. Even though the reported survival of DNA in dinosaur bones has now turned out to be erroneous, we know that DNA can persist for at least 40,000 years, as shown by the recent discovery of bits of DNA from the bones of a Neanderthal. It is possible that under special conditions, such as those found inside amber, DNA can survive for periods of tens of millions of years or more.

Despite many efforts, however, finding old RNA has proved difficult. Only one complete gene of the RNA from the virus responsible for the 1918 worldwide influenza pandemic has been resurrected from frozen bodies of its victims. It appears that RNA molecules cannot persist in the environment for periods longer than about a century. This instability of RNA has led scientists like Tomas Lindahl of the Imperial Cancer Research Fund in England to question whether the RNA world could ever have existed at all. In that chemically savage prebiotic world, RNA might have broken down as quickly as it formed.

If indeed it formed at all. How could RNA ever have appeared spontaneously? RNA contains only one of the two optical isomers of ribose (D-ribose) and uses only two of several possible pairs of chemical linkages between ribose and phosphate. It is very difficult to imagine how even the building blocks of RNA could have arisen by themselves, much less chains of RNA constructed from the building blocks.

RNA-world advocates point out that the multiplicative powers of replication are so great that reproduction could take place before the parent molecule decomposed, thus maintaining a critical number of self-replicating molecules. Moreover, the prebiotic world was a big, big place, with lots of molecules, a few of which might have been the building blocks of RNA.

But another group of scientists has taken a very different approach, one that ventures into utterly uncharted territory. Perhaps there was a world of genes before the appearance of RNA, a pre-RNA world.

Hints of such a world have been found in test-tube-evolution experiments. It turns out that such experiments need not be confined to RNA. Stephen W. Santoro and Gerald Joyce have shown that DNA can also be selected under test-tube conditions to catalyze some of the same reactions that ribozymes do.

The DNA used in these selections is not a classical double helical molecule of the kind that you see in textbook illustrations, because such a molecule would be inert and unable to perform any sort of catalytic function. Santoro and Joyce started with single-stranded DNA, a molecule very similar in its properties to RNA. Although their DNA enzymes were selected to do only one thing, which was to break RNA molecules at specific points, these results show that the type of sugar present in the nucleic acid backbone seems not to be critical for ribozyme-like catalysis.

THE FIRST PROTOBIONTS

Such experiments have led Joyce and others to ask whether other kinds of catalytic nucleic acid enzymes could have existed that were just as effective as RNA and DNA. Are the nucleobases present in DNA and RNA really the only ones that work? Is the phosphate in the nucleic acid backbone really necessary? How simple could such a molecule be? Of course, as Joyce has also pointed out, the molecule could not be too simple or it would be unable to function.

Such experiments have not yet been done, because the chemistry is daunting. To make ribozyme selection experiments work, researchers need special enzymes that convert RNA to DNA and back again. No equivalent enzymes exist that can act on other, simpler types of replicating molecules. Perhaps such enzymes existed once, in the pre-RNA world, but all the primitive organisms that made those enzymes have inconveniently gone extinct. Reconstructing such a world would require either the discovery of such organisms preserved today in some unusual ecological niche, or else a daunting amount of guesswork.

Peptide Nucleic Acids: A Glimpse of the First Genes?

Some progress has been made in finding candidates for these early genetic molecules. As Joyce has suggested, it might be possible to deconstruct the RNA molecule. Getting rid of the sugar would eliminate the optical isomer problem, and getting rid of the phosphate would eliminate the worry about how to bond it to a particular site on a sugar under plausible prebiotic conditions. Until recently there was no evidence that these things could be dispensed with and still permit the formation of a molecule that has retained most of the general characteristics of RNA. Now, however, discoveries in an entirely different field, gene therapeutics, make such a possibility more plausible. The candidate molecules are called peptide nucleic acids, or PNAs.

PNAs were originally synthesized by Peter Nielsen and his co-workers at the Panum Institute in Copenhagen, Denmark, in 1991. Their idea was to find molecules that looked like nucleic acids and that could bind to DNA and RNA, but that would do so irreversibly. Such molecules could be designed to bind themselves to pieces of the AIDS virus RNA, bringing replication of the virus to a screeching halt.

Like DNA and RNA, PNA consists of nucleobases held together in a chain, but they are joined together in a different manner. In the PNA molecule, both the sugar and the phosphate that are found in the back-

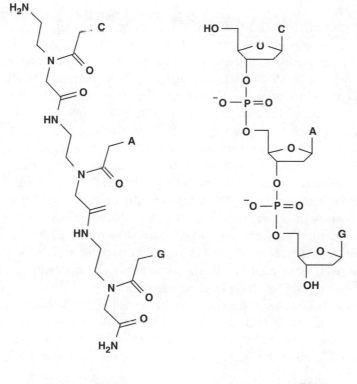

PNA   **DNA**

FIGURE 6.2   Diagrams of one strand of PNA and one strand of DNA, each having the base sequence CAG (C=cytosine; A=adenine; and G=guanine). The two molecules have the same overall shape, but the simple, wiggly backbone of PNA may have been much easier to synthesize in the prebiotic world.

bone of RNA and DNA have been jettisoned. Instead, the bases are joined through a molecule of acetic acid to a simple amino acid, which in turn is joined to other amino acid molecules to make the backbone of the polymer (Figure 6.2). The amino acid used for this backbone, 2-aminoethyl glycine (AEG), is not a natural amino acid. It shares an important property with the simplest of the natural amino acids, glycine, in that it does not have optical isomers. There is thus no handedness problem that needs to be overcome.

In effect, the monomeric units of PNA are joined together by peptide bonds, like those that link together amino acids in proteins. These are very stable bonds, which means that PNA is probably less likely than

THE FIRST PROTOBIONTS

RNA and DNA to break down over time—although few experiments have been carried out to verify if this is indeed the case.

The inventors of PNA immediately realized that this molecule might be a candidate for the first self-replicating molecule that preceded the RNA world. In 1993, Nielsen suggested the exciting possibility that there could have been a world of PNA-like molecules in the very distant past. Because of their overall similarity to RNA, such PNA-like molecules could well have given rise to the RNA world.

Because PNA molecules can have the same bases as RNA or DNA, they bond (base-pair) with other complementary PNA molecules, as well as with complementary molecules of both DNA and RNA, to form stable helical structures. These hybrid molecules are similar to those that form when complementary DNA and RNA molecules pair between themselves. Orgel and others have shown that PNA can also be used as a template for the synthesis of short stretches of DNA and RNA. It has even been possible, using some fancy chemistry, to construct short pieces of PNA using DNA as a template.

In short, PNA has properties very like those of the biologically active nucleic acids DNA and RNA. And the tough, simple, optically inactive PNA molecules probably could have arisen in the primordial soup more easily than the fragile, complicated, optically active RNA molecules.

There is compelling new evidence that both AEG and the acetic acid derivatives of nucleobases could have been present on the primitive Earth. Stanley Miller, working with graduate students Kevin Nelson and Matt Levy, has found that these molecules are produced in spark discharge reactions and during the polymerization of ammonium cyanide solutions. They conclude that PNA "is the first example of a plausible prebiotic genetic material."

A search for AEG and the acetic acid nucleobase derivatives in the Murchison meteorite is also under way. The detection of these key PNA components in meteorites would offer further evidence that they could have been readily made under prebiotic conditions and that they could have been present in the Earth's primordial soup.

There is every reason to suppose that joining the PNA monomeric units together might have been relatively easy under early Earth conditions. The linking, which involves the formation of peptide bonds, might have happened through the formation of AEG thioesters in a manner similar to that proposed by de Duve, Wächtershäuser, and Weber for making prebiotic peptides. An even simpler approach is being tested by Miller and his co-workers. Using solutions of AEG that have been concentrated by evaporation, as might have happened in lagoons or intertidal regions on the early Earth, they are testing whether AEG molecules will directly polymerize to form segments of the PNA backbone. So far the results look promising.

Could the first living entities have been nothing more than naked PNA molecules on a beach? None of the enzymes currently used in the test-tube-evolution experiments with RNA recognize PNA. But if this chemical difficulty could somehow be overcome, so that random sequences of PNA molecules could be subjected to test-tube evolution, then perhaps out of this collection of molecules one PNA molecule might be selected that could make copies of itself without assistance. If this self-replicating molecule turns out to be short and simple, then it might be very similar to the first self-replicating entity on Earth.

But if, as is more likely, a self-replicating PNA molecule turns out to be long and complicated, then the possibility that there was such a PNA gene sunning itself on an ancient beach will become vanishingly remote. Even though its chemistry is simpler than that of RNA, the PNA world would have been subjected to the same limitations as those that constrain the RNA world.

If the PNA world did exist on Earth, it is easy to see how it could have evolved into the RNA world. PNA molecules are very stable, which would have been an advantage in the hurly-burly world of the early Earth. But they are also highly reluctant to release their daughter molecules once they have been duplicated, which is of course the reason that PNA molecules were invented in the laboratory in the first place. Their replication in the world of early life would have been achingly slow, if it could have taken place at all.

In spite of this handicap, PNA molecules might have formed the basis for the first reasonably complicated life forms, albeit very slowly reproducing ones. Some of these in turn might have invented ways to make RNA building blocks. Any increase in the ease of replication would have given these protoorganisms an enormous advantage. And, since the information in PNA is coded in the same way as the information in RNA, all this information would have been transferred to the new rapid replicators.

A good analogy is the printing press. Once printing had been invented, the number of bibles in circulation skyrocketed, even though their content was the same as it had been when scribes had painfully copied them one at a time. In the same way, the new RNA organisms could easily have taken over from their poky PNA ancestors.

Self-Replicating Peptides?

This scenario of a PNA world may turn out, in the face of further experiments, to be wrong. Although many scientists working on the origin of life are homing in on PNA, they are also hedging their bets, for PNA is only one of many possible molecules that might have served in the pre-RNA world.

The researchers are right to be cautious. David Lee and his colleagues at the Scripps Research Institute in La Jolla, California, have now shown that autocatalytic capabilities are not confined to RNA or DNA or even PNA. They isolated a small peptide, part of a protein made by yeast, and showed that it could catalyze the joining together of two fragments of itself to make more copies of the complete peptide.

Here again, of course, the result is far from a completely self-replicating molecule. Such a molecule would have to start not with two pieces of itself but with a set of building blocks—in this case a collection of amino acids—and make a copy of itself from scratch. But Lee's remarkable experiment does show that yet another category of polymer can carry out the sort of reaction that might lead to a self-replicating system. And of course peptides—or at least proteinlike molecules—were probably present in the primordial soup.

Even if all the current excitement about PNAs or catalytic peptides turns out to be misplaced, one remarkable fact remains. The more we learn about the kinds of chemistry that could have taken place on the primitive Earth, the more likely it seems that some kind of complicated, information-bearing and energy-generating molecules could have appeared there. Prebiotic chemistry is turning out to be far more complex and versatile than could have been imagined when Miller did his first experiments. With discoveries such as these, and the many experiments that they open up, the gap between the prebiotic and the living world is slowly but surely closing.

Life Is in the Bag, or Is It?

The primordial soup and the slimy layers that accumulated on the rocks could have been very complex collections of substances indeed. They might have contained PNAs, peptides with limited autocatalytic capabilities, other sorts of molecules that had the properties of both peptides and nucleic acids, and even tiny fragments of RNA and similar molecules.

Manfred Eigen has suggested that even very simple molecules can cooperate in complex ways, passing information from one to another in what he calls *hypercycles*. These are chemical systems in which one set of catalytic molecules produces a product A, which plays an important role in a separate biochemical pathway. This pathway in turn produces a product B that might play an important role in the first pathway. Many such pathways, working together, would carry more genetic information than would be possible in a single pathway. Perhaps Eigen-like hypercycles could have been set up among this zoo of molecules, resulting in a cooperative collection of very different structures that, working together, were able to reproduce themselves.

Did genes come first, or did metabolism come first? Most likely, genes came shortly after at least a limited amount of organization had appeared in the molecular world. At its most elaborate, this organization would have provided the genes with some protection, along with a supply of building blocks, an energy supply, and "helper molecules" required for replication. Many workers in the field would agree that the naked gene is just too unlikely a concept. But scientists are divided over the question of just how organized the nonliving world could have been before genes could appear.

The idea that a protected place arose first, followed only later by the appearance of genetic material, is an old one among origin-of-life theorists. It harks back to Oparin and his coacervates, those cell-like structures surrounded by membranes that he hypothesized would be able to protect collections of delicate organic molecules from the ravages of the outside environment.

You will recall that an idea similar to Oparin's was proposed by Freeman Dyson in 1985, in his book *Origins of Life*. Dyson supposed that at some time near the beginning of the whole process, bags or globules of molecules could have arisen. These bags would have put in an appearance before the advent of genes as such, and were organized in such a way as to permit a primitive kind of self-replication. As a result, more-or-less identical populations of these bags of molecules would accumulate.

Dyson proposed that this might have happened because there would be a tendency for the molecules inside these bags to organize into metabolic pathways. The most successful of these pathways would be those that could make the largest number of pathway components. Each time the bags divided, they would contain larger and larger proportions of

THE FIRST PROTOBIONTS

these clever molecules. Eventually, the bags that replicated most easily and rapidly would predominate. Surprisingly, Dyson ruled out any role for Darwinian selection in his system.

Dyson's idea of little bags of molecules has been taken up and elaborated on by some other scientists, notably Günter Wächtershäuser in his metabolist model. Complex computer models exploring this type of scenario have been constructed by Doron Lancet of Israel's Weizmann Institute. Lancet has, however, not yet attempted to persuade molecules in the real world to undergo the same spontaneous organization that his imaginary molecules undergo in the computer.

Most scientists doubt that protoorganisms elaborate enough to have some properties of life could have arisen before genes. Leslie Orgel is particularly vocal in his opposition to these ideas, contending that there is no tendency for biochemical pathways, which consist of long sequences of chemical reactions that must take place in a highly specific order, to organize spontaneously in the test tube. To him, Dyson's bags seem even more unlikely than a pristine, naked gene. As a result, most scientists have tended to favor scenarios in which genes come first. And most research in this area has focused on how to make molecules in the test tube that can replicate themselves.

Let us suggest a middle road. As we pointed out in the last chapter, a simple hypothesis can aid in the difficult transition from random collections of molecules to the more orderly world of genes. The first collections of molecules from which life eventually arose need not have been organized into bags. Layers of molecules that shared the same property, such as sliminess or stickiness, would have accumulated in parts of the environment even if all the molecules that made them up, like Fox's proteinoids, were different from each other. These layers did not have the power to replicate, but they could attract molecules with similar properties from the primordial soup as they floated by. This enabled the layers to replenish the losses that resulted when some molecules were washed away or broken down chemically.

We might call this the spreading-slime hypothesis. A simulacrum of life is operating here. There is a crude kind of birth, in the sense that molecules with similar properties tend to accumulate in an environment that is favorable for them. And there is a crude kind of differential death, as some clusters of molecules are washed away or broken down, so that others with slightly different properties can take their place.

These layers might eventually have been able to carry out chemical modifications of the molecules making them up, and these modifications might have been fairly elaborate, though not as elaborate as the ones that Dyson postulated for his bags. Of course, the layers would still have been made up of crazy quilts of molecules, with all kinds of different processes taking place inside them, but they would have had one thing in common. They would, as we will see, have been good to "eat" by the first living entities.

Darwin Takes Charge

Freeman Dyson made another suggestion in his book. He supposed that the first self-replicating structures might have been tiny primitive "viruses" that preyed on his bags of molecules. The idea is very sensible, but the term *virus* is perhaps an unfortunate one. Present-day viruses can only multiply inside the cells of their hosts, and most viruses are highly specialized to live inside a particular type of host cell. Let us instead call these self-replicating structures the first *protobionts,* a term used for the first living structures but which we can define here as collections of molecules that carried at least some genetic information and that could multiply inside the complex, slimy layers of molecules that had accumulated on the rocks.

What might these first protobionts have been like? We agree with Dyson (and with Eigen) that these first self-replicating entities must have been very simple. Nonetheless, they probably had several components. One part would have been a fragment of genetic material, made up of a short piece of nucleic acid or a similar molecule. This little fragment did not need to be very big or elaborate. It could have arisen anywhere, but it most probably appeared somewhere in the slimy layer itself.

Other parts of the protobiont were molecules, again initially very simple, that aided this bit of genetic material in replicating itself. Some of these molecules provided energy for the process, by mechanisms that we will examine in the next chapter.

Taken together, these molecules made up the first truly self-replicating entities. And, just as present-day viruses can only replicate themselves inside their host cells, the first protobionts could only have replicated themselves within the "host" slime layer. Even if only a few small parts of the host layer were suitable, this might have been enough for very slow replication of these protobionts to take place.

How likely is this? After all, no sign of anything as elaborate as nucleic acids has yet been found in any of the prebiotic synthesis experiments or in meteorites. But remember two things. First, individual molecules are very small, and as we have discussed earlier, even if they were present in dilute amounts, there would still have been millions of them. The soup must have been full of a bewildering variety of such molecules—conceivably even including a few short bits of nucleic acid that had polymerized on some particularly favorable mineral surface. Second, we must not forget the power of replication. It would take only one such molecule, if it found itself as part of a protobiont on a favorable host substrate, to multiply into many.

These first protobionts might initially have coded for nothing except a determined ability to make copies of themselves inside the safe and resource-rich environment of the slimy layer. But they could go on replicating because, even though they would soon exhaust the supply of building blocks, the layer would attract new ones. Remember that the layer has by that time been selected by proto-Darwinian evolution to be very good at replenishing itself!

The building blocks would have been of many different types, and as a result, the copies that the protobiont made of itself would have been by no means exact copies—in effect, the mutation rate was extremely high. Before long, some versions of these replicating molecules would begin to affect their immediate environment. They might, simply by accumulating in large numbers, alter the ability of the host layer to repair itself. Their presence might be enough to attract larger numbers of suitable building blocks from the passing throng.

It is here that Darwin really begins to take charge. Once a self-replicating molecule begins to modify its environment, even if it does not initially code for any aspect of that environment, it will begin to evolve to compensate for the changes it is causing. There would be strong selection for some of the self-replicating molecules to code for protein molecules that would aid the ability of this environment to repair itself and increase in size.

Some of these proteins might have had numerous attachment sites for sugars, so that as they multiplied, they would repair the slimy protective coat on the surface of the layer. Others might have attracted and bound the building blocks needed for the protobiont to replicate itself, increasing the rate at which this simple structure could replicate.

New strains of protobionts, better able to invade new areas and to do so without depleting them of resources immediately, would have been selected for. And before long, some of them would code for host mole-

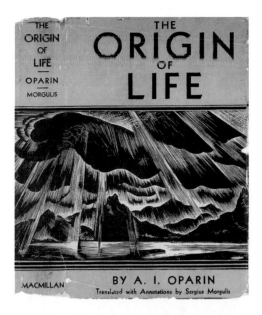

PLATE 1. The cover of the 1938 English translation of Oparin's book Origin of Life, published by Macmillan, New York.

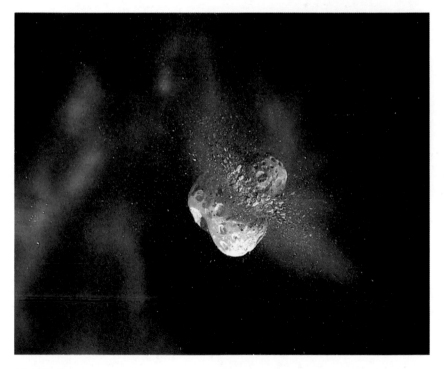

PLATE 2. Collision between two planetesimals during the formation of the Solar System. (Painting by Bill Hartmann, Planetary Sciences Institute, University of Arizona)

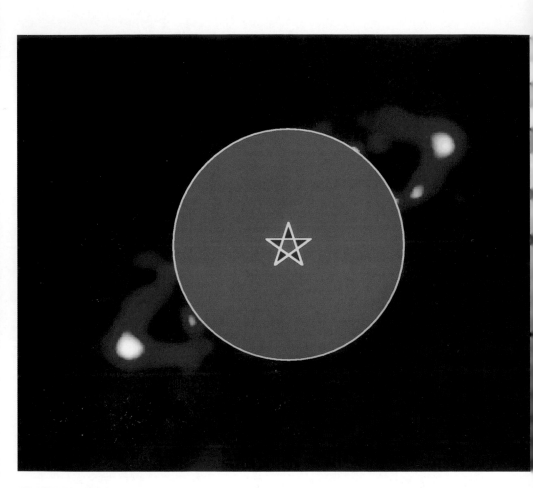

PLATE 3. Evidence for planets in the process of formation is seen as a tightly confined ring of ciricumstellar debris, seen in reflected light, around the young (~ 10 million year old) main-sequence star HR 4796A. (Courtesy of Dr. Glenn Schneider and NASA).

PLATE 4.   The Moon soon after its formation, with the still largely molten Earth in the foreground. (Painting by Ron Miller, courtesy Bill Hartmann.)

PLATE 5.   The early Earth-Moon system, showing the disk of material from which the Moon was formed. In the picture, the Moon is casting its huge shadow on the nearby Earth. (Painting by Bill Hartmann.)

PLATE 6.   A volcanic eruption on the early Earth, showing the dark smoggy atmosphere and tremendous lightning displays that accompanied it. (Painting by Bill Hartmann.)

PLATE 7. One of the first islands to form on the Earth. The picture shows a bright sunny day, but the atmosphere is likely to have been dark and smoggy. (Painting by Bill Hartmann.)

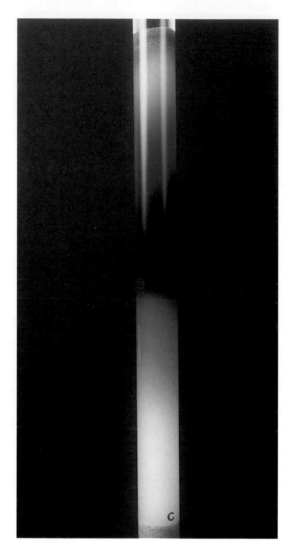

PLATE 8. Picture showing the chromatographic separation of the colors present in the water-soluble black ink of a felt-tip pen. The ink was mixed with sand and then placed on top of a column of fine-grain silica. Water dripping slowly through the column separated the colored components. Natural chromatographic processes would have been important in separating the components of the primordial soup. (Scripps Institution of Oceanography)

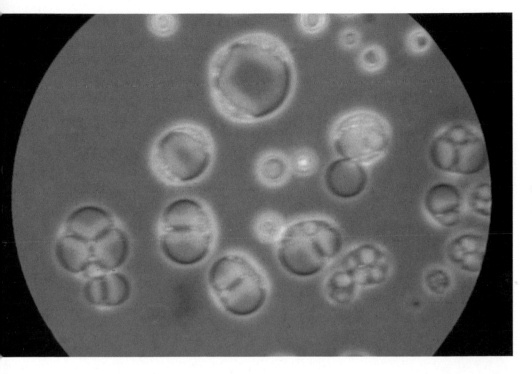

PLATE 9.   Globules formed in an extract of oily materials from the Murchison meteorite. If there was water on the parent body of this meteorite, then similar globules would have formed there four and a half billion years ago, at the dawn of the Solar System. (Courtesy David Deamer)

PLATE 10. A depiction of the Huygens spacecraft as it might look when it lands on Titan. (Courtesy of ESA)

cules that would surround and protect them. This would give them a safe and growing place to live.

Given a host layer of molecules capable of self-replenishment, any protobiont-like molecule that could replicate and exploit that layer would be selected for—and this, primitive as it was, would have represented the first stirrings of life. Now, with replicated information being passed from one generation to the next, the full range of Darwinian evolution could be brought to bear. Darwin Lite would have been replaced by Darwin for real.

Thus, although Dyson's first idea of self-replicating, gene-free bags of molecules is very unlikely, his second idea, that the first genetic material might have been a tiny, viruslike fragment, is a powerful one worth exploring in the laboratory. We find this possibility particularly appealing because it provides a way for genes to start out extremely simple and gradually become more complicated with time.

There is no obvious way by which the many different genes needed for life as we know it could have arisen at once. A great many genes are needed to code for the information required to make even the simplest cell alive today. The simplest independently living organism that we know of at present, a tiny bacterium called *Mycoplasma*, has fewer than five hundred genes, but it still needs over half a million bases of DNA to code for these genes. The AIDS virus, in contrast, has only ten genes, but it cannot live independently of its host cells.

The viruslike protobionts at the beginnings of life must have had few genes—or even as we suggested none at all—but nonetheless they must have been able to multiply in a highly diverse chemical environment. No such living entities can be found today. The viruses of today are all extremely specialized to live only on certain types of host cells. The kind of world that made those highly versatile protobionts has long since vanished.

How big could those very earliest pieces of genetic information have been? Far less extensive than half a million bases, certainly. And it is unlikely that the earliest replicating protobionts would have been supplied with a set of building blocks and energy-rich molecules that were all the same—a nice, clean supply of ATP and so on. They would have had to replicate as best they could, using a dirty collection of molecules of approximately the right size and properties. Each resulting molecule was unlikely to have been exactly like its parent.

In these chapters, we have gone as far as is currently possible to go using the bottom-up approach. To progress further, new experiments are

THE FIRST PROTOBIONTS

needed that replicate the conditions on the early Earth with ever-greater accuracy. Such explorations of "dirty" chemistry, coupled with new techniques to detect replicating molecules, will allow us to explore the conditions needed to produce primitive protobionts. But there is another way to approach the origin of life, and that is from the top down.

FROM TOP TO TOE

One step above the sublime, makes the ridiculous, and one step above the ridiculous, makes the sublime again.

Thomas Paine, The Age of Reason *(1794)*

THROUGHOUT THE MODERN LIVING WORLD, THE CURRENT genetic code is almost universal. There are a few minor dialect variants, but in general when you read the sequence of DNA of any organism, you will know exactly what proteins it makes.

It is now clear that this complex code is no evolutionary accident, but began as something simpler and grew more complicated. We know this in part because of work by Steve Freeland and Laurence Hurst of the University of Bath. They asked a simple question: What would happen if you rearranged the code? Would it be as good as the original code?

To understand what they did, we must know something about the code itself. The sequence of bases arranged along a strand of DNA is read in groups of three by a form of RNA called messenger RNA (mRNA). Each group of three, called a *codon,* specifies which amino acid will be incorporated into a growing peptide chain.

At present (although this may not have been true when life began) there are four bases in RNA: uracil, adenine, cytosine, and guanine, represented as U, A, C, and G. In DNA, the U is replaced by a T (thymine). This means that the DNA genetic code transcribed by mRNA consists of sixty-four (4 x 4 x 4) codons, because there are sixty-four possible ways in which the four RNA bases can be arranged in groups of three. Three of these codons specify stop signals in the peptide synthesis process, and the remaining sixty-one code for specific amino acids. When the code of a gene is read in its entirety by the complex machinery of the cell, the result is a protein in which the sequence of amino acids precisely reflects the sequence of codons in the gene (Figure 7.1).

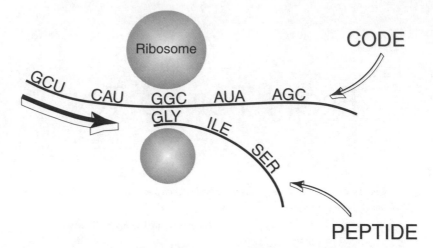

FIGURE 7.1 Ribosomes are like little printing presses. The code for a protein is in the form of a messenger RNA molecule that is transcribed from the original DNA gene, and fed into the ribosome (big arrow). A growing peptide chain that corresponds to the code emerges from the ribosome. The code is read in groups of three letters, called codons.

Because only twenty amino acids are used as building blocks for proteins, most amino acids are specified by more than one codon. This property varies from one amino acid to another. The amino acid tryptophan, for example, is specified by only one codon, but the amino acid arginine is specified by six. Thus, some mutations that change bases will substitute one amino acid for another, and some will leave the resulting protein unchanged.

Using the computer, Freeland and Hurst constructed large numbers of imaginary genetic codes. They took a given set of codons, such as the ones that specify arginine, and assigned them to another amino acid at random, at the same time assigning arginine to a different set of codons. They did this repeatedly, and after they had finished shuffling all the assignments around, they then asked whether this new code would be more or less likely than the real one to give rise to mutations that would change critical amino acids, and whether it would be more or less efficient at translating proteins. They found that only about one in a million of these new codes was as good as the real code—the rest were worse, sometimes dramatically so.

This means that the present-day genetic code must have been refined extensively through natural selection, or else we would have ended up with one of the many poorer codes. Freeland and Hurst realized that this selection must have taken place a long time ago, before organisms became too complicated and genes became too numerous. By now, the cost of changing the genetic code in a huge and complex organism like a human is overwhelmingly large—a single change in the code, no matter how minor, would result in so many harmful mistakes scattered around our tens of thousands of genes that we would surely die. Therefore, the most likely time during which all this selection took place was when the code itself was just starting to grow in complexity.

The code must have started out very simple, and may indeed have been rather imprecise. It became more refined and precise over time. But how simple, and how imprecise, was it at the beginning?

Suggestions have been made that the code started out not as a three-letter code but in a much simpler form, and that perhaps not all four bases were used at the outset. Manfred Eigen has suggested a two-letter, two-base code that could have coded for four amino acids, not the twenty of the present code. But there are other ways to strip the code down to its essentials.

In 1966, Walter Fitch (now at the University of California at Irvine) took the idea of a simple code to its extreme. He imagined a code that initially did not make any distinction at all among the amino acids! Then, through a process that he called ambiguity reduction, the code gradually came to specify more and more of the properties that distinguish the various amino acids.

We can follow his logic and see how such ambiguity reduction could have taken place. One property of the code is particularly intriguing and gives us a clue to how it may have started out. When the code is arranged in a table, as you see in Figure 7.2, a remarkable feature of its organization leaps out at you. If you draw a line running vertically down the middle of the table, you find that most amino acids on the left have hydrocarbon-like (nonpolar) side chains and tend not to interact with water. Most on the right have charged, or water-loving (polar), side chains, and are much more chemically active than those on the left.

You can see from the table that it is the middle base of each codon that is the cause of this particular distinction. The bases that occupy the first or third positions in the codon are largely immaterial as far as this important property is concerned.

Suppose that in the beginning, the code used only two of the four bases. Suppose further that the codon was three bases long, as it is today, perhaps because this was dictated by the various sizes of the

First Base In Codon	Second Base In Codon				Third Base
	U	C	A	G	
U	Phenylalanine	*Serine*	Tyrosine	**Cysteine**	U
	Phenylalanine	*Serine*	Tyrosine	**Cysteine**	C
	Leucine	*Serine*	Termination	Termination	A
	Leucine	*Serine*	Termination	Tryptophan	G
C	Leucine	Proline	**Histidine**	Arginine	U
	Leucine	Proline	**Histidine**	Arginine	C
	Leucine	Proline	**Glutamine**	Arginine	A
	Leucine	Proline	**Glutamine**	Arginine	G
A	Isoleucine	*Threonine*	**Asparagine**	Serine	U
	Isoleucine	*Threonine*	**Asparagine**	Serine	C
	Isoleucine	*Threonine*	**Lysine**	Arginine	A
	Methionine	*Threonine*	**Lysine**	Arginine	G
G	Valine	Alanine	**Aspartic Acid**	Glycine	U
	Valine	Alanine	**Aspartic Acid**	Glycine	C
	Valine	Alanine	**Glutamic Acid**	Glycine	A
	Valine	Alanine	**Glutamic Acid**	Glycine	G

FIGURE 7.2   The way the genetic code splits according to the second base in the codon. The majority of amino acids to the left of the heavy vertical line are water-hating. The only exceptions are serine and threonine, shown in italics. Most of those to the right are water-loving. The most chemically active of the water-loving molecules are shown in bold.

molecules involved. But in its earliest form, only the middle base mattered. Thus, just as it does today, a U in the middle position might specify water-hating amino acids and an A at that same position might specify water-loving, more reactive amino acids. And because A and U pair, this most primitive of genes could have replicated in the same way as genes do today.

Such a code, read codon by codon, would have produced a string of binary information—in effect a string of ones and zeros. U's in the second codon position would have coded for one kind of amino acid, and A's would have coded for another. But how could such a simple code, carrying such a minimal amount of information, possibly have resulted in proteins that possessed the diversity of different properties needed for emerging life?

It might have done so if U's coded for not just one kind of amino acid, but rather for a set of amino acids with similar properties, while A's coded for a different set of amino acids having different properties. This could have happened if the early code was not only primitive but also nonchalant about its job. Perhaps it only made a distinction between water-loving and water-hating, and as long as an amino acid fell into the right one of these two categories, it would cheerfully paste that amino acid into the growing protein chain.

This primitive sort of gene would be like a jewelry maker who uses an eclectic assortment of beads of various sizes to make necklaces. There might be only one rule, for example, that there must be larger beads in the middle of the necklace and smaller ones at the ends. Although each of the resulting necklaces would be different, they would at least share this property.

Given a sufficient level of nonchalance, exactly which amino acid ended up at a particular point in any given protein molecule would have been a matter of chance, but the code would at least have dictated whether the amino acid was water-loving or water-hating. And although the next protein that was made from that code would not have had the same amino acids in the same positions, at least it would have had the same distribution of water-loving and water-hating regions along its length.

Proteins made in such a way would all be different from each other. This might not seem much good to us, because we are used to the precision of the present-day molecular world, in which billions of identical copies of a protein can be made by our cells. But sets of proteins that shared only one or a few properties might have been useful catalysts in a very primitive world.

The distinction between water-loving and water-hating would have been important, because the proteins' water-hating regions would have been attracted to fatty membranes, and their water-loving regions would have projected out and been available for catalysis. As we will see shortly, proteins that got stuck in membranes, like flies on flypaper, might have been extremely important for early life.

As Fitch pointed out, such a relaxed code could have used many of the amino acids that are not currently used in proteins but that were found in abundance in the primordial soup. All that would have been required was that the code recognized their water-loving or water-hating properties. Alpha-amino isobutyric acid is an amino acid without handedness, and it is similar in shape to both D- and L-alanine. It is made in large quantities in primordial soup experiments and is one of the dominant amino acids found in the Murchison meteorite. Perhaps it and amino acids with similar properties served as temporary proxies for the L–amino acids that eventually took over when the decision that favored L–amino acids was made.

The code itself need not have been carried by RNA, and it need not have always used A and U (or G and C). If the coding message, too, was imprecisely replicated, this would increase the resulting variety of proteins.

The first of the protobionts that actually carried any genetic information must have done so using just such a sloppy code. But as greater demands were placed on the protobionts, the code would soon have begun to change, and ambiguities would have been reduced. More bases would have been incorporated, perhaps allowing less stable bases like cytosine to take part. The first position in the codon, and finally the third position, would have begun to play a role. The nonchalant nature of the first code, which did not at first care exactly which amino acid it coded for as long as it was water-loving or water-hating, would have been replaced by a code of greater and greater precision. Fitch has found evidence, from the way the rest of the protein machinery evolved, that this is probably what happened. He has also found in the code several traces of further ambiguity reduction.

Such an initially sloppy code also gets around the problem posed by Eigen, that a long piece of code would have accumulated too many mutations to be useful. Eigen assumed that the information carried by the code must be very precise and that mistakes are a matter of life and death. If only a small amount of information was carried by this early code, and the information was of rather low quality, then those early protobionts could have survived a very high rate of mutation and still have been able to pass their information on to the next generation.

Nonetheless, Eigen was right when he pointed out that genes could not have been very large. Most of these evolutionary changes in the genetic code could only have taken place if these early genes carried relatively small amounts of information. By the time genes had become long and numerous, all the really dramatic changes that have shaped our "one in a million" genetic code must have taken place.

Once genes appeared, full-bore Darwinism ruled. Genes could now be acted on by natural selection, which up until this time could only act in the crudest way on the molecules of the prebiotic soup or the spreading slime. Protobionts began to replicate in numbers and to mutate, evolve, and change the world they inhabited.

But as soon as this happened, a new problem loomed. The bounty of the primordial soup could not go on forever. The planet was changing. Vents were destroying organic molecules more rapidly than they could be replenished. Something dramatic had to happen, and soon, or this nascent life would expire from lack of nourishment.

The Origin of Energy-Utilizing Systems

Hell is usually described in terms of fire and brimstone, the latter an old term for elemental sulfur. The description must have arisen from observations of that most obvious manifestation of hell on Earth, volcanic eruptions. Ironically, life most likely began, not in a heavenly Garden of Eden, but in some unpleasant approximation of the other place. And it is surely not a coincidence that compounds of sulfur and iron, produced so plentifully by volcanic eruptions, are also intimately embedded in the proteins that form part of the energy-producing machinery of our own cells. We all have a faint echo of the infernal beginnings of life in our cells.

One place in which this echo can still be found is inside our mitochondria. Mitochondria are little globules surrounded by membranes and present in large numbers inside most of our cells. They are absolutely essential to our existence, for they are our main source of ATP and are also involved in the manufacture of energy-rich fats and many other important compounds. They have the ability to use oxygen directly to oxidize—in effect to burn—simple organic compounds, extracting a huge abundance of energy from them.

Our mitochondria are in fact highly modified structures that have had a long evolutionary history themselves. Their remote ancestors were once free-living bacteria that invaded the cells of our ancestors. In other words, our cells contain the much-simplified descendants of ancient bacteria. Any discomfort you may feel about this should be tempered with the knowledge that we have benefited immensely from that ancient invasion.

One of the reasons we know about that invasion is that versions of the same energy-generating system that is found in mitochondria can also be found in bacteria. The entire mitochondrion is involved in ATP generation, and in bacteria the ATP-manufacturing system involves the entire cell.

The story of how this energy-producing machinery evolved is a fascinating one and is essential to our understanding of the origin of life. By taking the top-down approach and following the tale back in time, we can catch a glimpse of how the first protobionts obtained the energy they needed to make copies of themselves.

The most important character in this story is not a scientist directly involved in the quest for the origin of life. It is English chemist Peter Mitchell (1920–1992), who first understood how the living cell makes energy. His revolutionary ideas were initially rejected by the scientific community.

Peter Mitchell's Battle with Conventional Wisdom

The mechanisms by which mitochondria work eluded biochemists for many years. In fact, arguments over this problem led to one of the great scientific controversies of the twentieth century. The majority of biochemists were convinced that ATP in mitochondria is produced by the activity of ordinary enzymes. Peter Mitchell stood alone against this view for many years. He proposed a startlingly different idea, that it is the very structure of the mitochondria that is essential to the manufacture of ATP.

Mitchell was educated at Cambridge University and quickly acquired a reputation as a maverick. In 1961, twenty years after graduating, he eventually became a Reader at the University of Edinburgh, but soon after his appointment, the strain of his battles with the establishment took its toll. While recovering from severe gastric ulcers in 1963, he decided to move completely outside that establishment.

He discovered an extremely run-down Regency-fronted mansion in a remote part of Cornwall, renovated it, and established a tiny research foundation. Here, with a small group of fellow researchers, and supported by donations, he carried out most of the experiments that overturned the accepted view of ATP synthesis. Eventually, through a series of brilliant demonstrations of how mitochondria really worked, Mitchell's view prevailed. He was awarded the Nobel prize in 1978— and that year, unlike most years, nobody else shared the prize.

Why the controversy? Mitchell had introduced a daring new concept into the orderly world of biochemistry. In that world, as it had been explored for decades by biochemists, chemical reactions are carried out by enzymes that are entirely self-sufficient. They do not need the assistance of any larger entities. Enzymes should be able to carry out their tasks anywhere in the cell, because they can create their own minienvironment that allows specific chemical reactions to take place.

Enzymes are marvelously sophisticated catalytic molecules, and they operate like tiny molecular jigs. In very precise ways, they seize and grip the small molecules that are to take part in the reaction. This positioning greatly increases the likelihood that the reaction will proceed, usually through a series of discrete and detectable steps that yield unstable intermediate molecules. The enzyme keeps a firm grip on these intermediates until the reaction has been completed.

Scientists were puzzled, then, that no sign of any obvious intermediate molecule that would be expected to lead to ATP had ever been found in the mitochondria. This was a great mystery, to which traditional biochemistry had no answer. Mitchell was also well aware that as soon as a mitochondrial membrane is broken or damaged, the production of ATP ceases—even though all the proteins involved are still intact. His first profound contribution was the realization that the membranes that surround the mitochondria are an essential part of the ATP-generating process.

We now know that the mitochondrial membrane, and indeed all other biological membranes, are extremely important to life. About a third of any living cell is made up of membranes, and these membranes divide the cell up into all sorts of compartments. Membranes are made up of fatty-acid molecules that spontaneously adhere to each other. One end of each of these molecules is water-loving, and the other is water-hating. The water-hating ends readily attract each other to form sheets, which leaves the water-loving ends to face both the water-filled outside world and to the water-filled cell. The result is a wall-like structure, the inside of which excludes water (Figure 7.3).

Many proteins, too, have water-loving and water-hating regions, and the water-hating parts can literally dissolve into the interior of the membrane. When this happens, the protein is caught in the membrane like a fly in amber, often with the water-loving parts sticking out of the two sides of the membrane. You will remember that the most primitive distinction that the early genetic code must have made was between water-loving and water-hating amino acids, and this distinction strongly suggests that the interaction between proteins and membranes must be very old indeed.

Mitchell and others showed that mitochondrial membranes, along with the protein molecules floating in them, work together to set up conditions that can drive an ATP-generating system. Some proteins in the mitochondrial membrane can pick up electrons from energy-rich compounds. These electrons are then passed from one of the floating protein molecules to the next, like the legendary Chicago Cubs double plays in which the ball was hurled from Tinker to Evers to Chance.

Iron and sulfur play important roles in all of this. These elements, buried inside the floating proteins, are echoes of a very ancient world in which pyrites must have been important catalysts. By happy chance, these iron–sulfur complexes are excellent at picking up electrons and releasing them again.

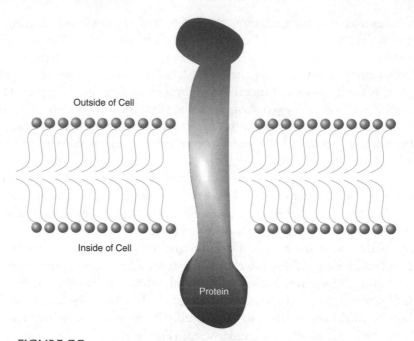

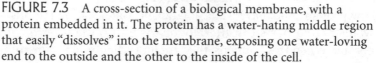

FIGURE 7.3 A cross-section of a biological membrane, with a protein embedded in it. The protein has a water-hating middle region that easily "dissolves" into the membrane, exposing one water-loving end to the outside and the other to the inside of the cell.

As the electrons are passed from one protein to the next, each protein becomes briefly charged with energy. Mitchell's most important contribution was to show that this energy is used to pump protons from the inside of the mitochondrion to the outside. Protons are positively charged hydrogen nuclei (chemists give them the symbol H^+) that are abundant in biological solutions. Mitchell demonstrated that they can do chemical work just as easily as electrons can.

The most important work that these protons carry out is the manufacture of ATP. We saw earlier how ATP is fundamental to biochemical processes in all living cells. And we remarked at the time how it is surely not a coincidence that it contains adenine, which John Oró discovered can be so easily synthesized from cyanide.

You will remember, however, that ATP is a much more complicated molecule than adenine. Besides adenine, ATP also contains ribose, the five-carbon sugar also found in RNA. In the synthesis of ATP, a chain of

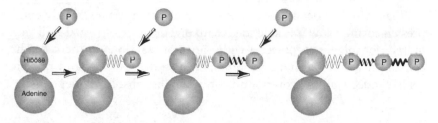

FIGURE 7.4   The stepwise addition of phosphates to adenosine becomes progressively more difficult and energy-demanding. The result is an energy-packed molecule that can perform many different tasks in the living cell.

three phosphate molecules is attached to the sugar, each joined to the next by increasingly energy-rich bonds.

This chain is built in a series of steps in our bodies, steps that are easy at first and become progressively more difficult (Figure 7.4). It is not too energy-demanding to start with adenosine and add a phosphoric acid to make adenosine monophosphate. But it is considerably more difficult to add a second phosphoric acid to make adenosine diphosphate. It is most difficult of all to add that final phosphoric acid and make adenosine triphosphate or ATP. The addition of these phosphoric acids, like the joining together of amino acids to form peptides, takes place through a dehydration reaction (Figure 7.5).

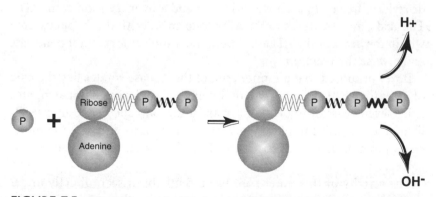

FIGURE 7.5   Phosphates are added to make ATP by an energy-demanding process in which water, in the form of a proton (H+) and a hydroxyl ion (OH-), is removed. The pH gradient across the membrane sends the two ions in opposite directions.

FROM TOP TO TOE

When ATP has been made, it is as full of energy as a strongly compressed spring. Now it is able to give up one or more of its painfully acquired phosphates to power a wide range of reactions in the cell. The cell can use ATP to make proteins, carbohydrates, fats, and nucleic acids—indeed, all the compounds on which life absolutely depends.

ATPase: The Core of Life

The result of all that pumping of protons is a dramatic difference in pH, or acidity, between the inside and the outside of the fatty membrane that surrounds the mitochondrion. The outside is about fifteen times as acidic as the inside.

Once a pH difference across the membrane is established, the protons will want to flow back again and equalize the pH on the two sides of the membrane. In the process, they can do useful work. Another kind of protein floating in the membrane can take advantage of this and can use the protons to make ATP. This molecule—or rather collection of molecules—is called ATPase, and it lies at the very heart of life itself.

Even though ATPase has been studied intensely for decades, the exact way in which it works is just beginning to be understood. The 1997 Nobel prize was given in part to Paul Boyer of the University of California at Los Angeles and John Walker of Cambridge University, who have come closer than anyone else to the heart of this fundamental question.

The ATPase lets protons return to the other side of the membrane through a narrow opening. It uses the resulting energy to tear a hydroxyl ion free from a phosphoric acid and a hydrogen ion from ADP. The acid can then attach to the ADP to form ATP. All this happens deep within the heart of the ATPase, where water molecules cannot penetrate and reverse the reaction.

Boyer proposed that an inner core of the ATPase rotates like the core of a Mazda rotary engine, briefly opening and closing compartments. After seven years of effort, Walker and his group succeeded in crystallizing the enzyme complex, which allowed them to visualize the molecule's three-dimensional structure. They found that Boyer's scheme is correct.

The activity of this inner core has recently been seen directly under the microscope. When researchers have attached a fluorescent tag to the core, they have been able to watch the tag whirl around and around. The ATPase complex, like some giant, elaborate Robocop, thrums with energy and spews out ATP at a tremendous rate.

The essential point, as Peter Mitchell had realized, is that the ATPase is very different from an ordinary enzyme, for it cannot do its job if the membrane is broken or if the ATPase has been isolated and separated from the membrane. Even though all the parts of the ATPase are still intact, there will be no flow of protons to power the rotation of the inner core.

In our mitochondria, the entire structure becomes an ATP generator. And in bacterial cells, which are surrounded by a membrane that has all these molecules embedded in it, the whole cell is an ATP generator. The membrane and the molecules floating within it are one seamless ATP factory, which means that they must have evolved together. It is impossible to imagine the ATPase, the electron-transport proteins, and the membrane of which they are such an integral part evolving separately from each other and then coming together. Right from the beginning, the simpler ancestors of these various proteins must always have been together, embedded in some primitive version of today's more sophisticated membranes.

The Ancestor of the Robocop Enzyme

How could something so complicated have evolved? Like the process of Darwinian evolution and the evolution of the genetic code, it must have started out far more simply. So another way of phrasing the question is to ask how simple could such a collection of molecules be and still have the capability of generating ATP or an equivalent high-energy molecule.

Some of the most intriguing experiments to investigate this question have been conducted by David Deamer of the University of California at Santa Cruz. He has argued, as we have done here, that one of the very first events in the evolution of life must have been the appearance of mechanisms for extracting energy from the environment. Such systems are likely to predate, or at the very least accompany, the first appearance of genes.

The fundamental question that Deamer poses is whether there is something about the chemistry of the molecules present on the early Earth that could have led to the formation of crude energy-extraction systems. He took two ingenious approaches to try to answer this question.

The first was to examine our old friend, the Murchison meteorite, to see whether it might contain molecules that could form some kind of

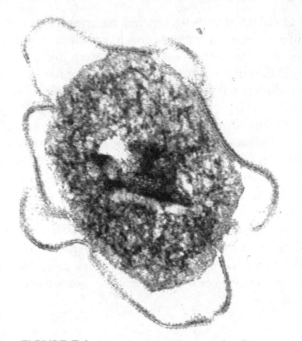

FIGURE 7.6 An electron micrograph of a coacervate-like structure formed from an extract of the Murchison meteorite. Note the distinctly bilayered membrane around the outside that is reminiscent of the bilayered membrane found around living cells today. Plate 9 shows a color picture of these little globules, seen under the light microscope. (Courtesy David Deamer)

membranes, and whether these membranes might have had at least some of the properties of those in today's living cells. He took some precious fragments of the meteorite, ground them up, and shook the fragments together with solvents that would dissolve water-insoluble materials like fats and oils. The small amount of material that he obtained was made up of a collection of molecules far too diverse for him to analyze, but he found that some fractions of it, dissolved in water, formed little globules that looked for all the world like Oparin's coacervates. When he examined ultrathin slices of the globules under an electron microscope, he found that some of the membranes surrounding these little globules were made up of a double layer, like biological membranes (Figure 7.6).

Did little coacervate-like globules form in tiny pockets of "ocean" that existed briefly inside Murchison's parent asteroid? And what else happened before the asteroid was smashed to smithereens early in the history of the Solar System?

There were too few of these little globules to do experiments with. But Deamer knew that at least some components of the globules were polycyclic aromatic hydrocarbons (PAHs). What would happen if such molecules were sequestered inside membranes?

To find out, he took fat molecules of the type that form present-day membranes, and shook them up with PAHs. Little globules surrounded by a fatty layer formed, and, like the proteins of the mitochondrial membrane, the PAHs were sometimes actually embedded in the layer.

The next step yielded a remarkable result. When Deamer shone light on his suspension of little globules and measured the pH inside and outside them, he found that the pH was different! This was not a directional pump like the one found in the mitochondria, for he discovered that the protons were moving in both directions through the membrane. Instead, the pH difference appeared because protons tended to get trapped inside the globules and to diffuse away outside. Nonetheless, he had managed to replicate one of the basic phenomena that underlie the generation of energy-producing systems, a light-driven transfer of energy, starting with very simple materials of the kind that might have been found on the primitive Earth.

Today, light-driven energy transfers play a central role in photosynthesis. We are starting to learn about the origins of photosynthesis by studying photosynthetic bacteria rather than green plants. Today, many different bacteria can carry out photosynthesis, and many of them, unlike green plants, do not make oxygen in the process. Through a study of these primitive types of photosynthesis, we will learn more about the sources of energy in the early world.

A century and a half ago, long before the idea of a primitive soup, German zoologist Ernst Haeckel proposed that the first living organisms were photosynthetic. Oparin dismissed this idea, insisting that the first organisms must have obtained their energy by dining on the primitive soup. Perhaps, as with so many origin of life theories, Haeckel's ideas will soon be dusted off and reexamined!

The story of how the first protobionts obtained their energy is now beginning to take shape. Oparin realized the importance of membrane-bound aggregates of molecules, Mitchell demonstrated the existence of their remote descendants in present-day cells, and now Deamer is

FROM TOP TO TOE

pointing the way toward creating membrane-bound aggregates with new properties in the laboratory.

Such coacervate-like globules must often have been caught up and embedded in the spreading slime, and some proteinlike and PAH molecules that had water-hating middle regions must have been trapped in their membranes. Among the billions of these evanescent globules, a few might by chance have been able to phosphorylate molecules like adenosine, leading to an accumulation of charged molecules available for the synthesis of short bits of nucleic acid.

It is unlikely that these globules were self-replicating, like Dyson's little bags, but some must have been able to sequester, protect, and nurture the first primitive living entities, the protobionts. Within and among these small and fleeting collections of globules, hiding beneath a protective outer coating of slime on wave-scoured rocks, the next step in the origin of life must have taken place.

JOURNEY TO THE CENTER OF THE EARTH

I had a dream, which was not all a dream.
The bright Sun was extinguish'd, and the stars
Did wander darkling in the eternal space,
Rayless, and pathless, and the icy earth
Swung blind and blackening in the moonless air . . .
The winds were wither'd in the stagnant air,
And the clouds perish'd; Darkness had no need
Of aid from them—She was the Universe.

       George Gordon, Lord Byron, Darkness *(1816)*

LIFE MOST LIKELY APPEARED ON THE EARLY EARTH'S BEACHES and tide pools and the surrounding rocks. These regions were washed with waves and tides and provided Darwinian challenges from the very beginning. But we must remember that then, as now, the Earth was a huge place with a wild and varied terrain. We cannot rule out the possibility that life could have appeared in places other than the oceans, and indeed even perhaps in the depths of the rocky crust. But it is far more likely that rapidly changing environments and the presence of energy sources such as lightning and sunlight would have sped up the processes associated with the origin of life. As a result, even if life was able to begin in places other than the Earth's surface, its nascent beginnings would have been overwhelmed by the swiftly evolving life-forms that gestated in the intertidal regions.

Now we are discovering that there may be other places in our Solar System where life might have appeared, even though there were probably no tides and beaches to help the process along. Could life have arisen on the tideless beaches of early Mars, or in the depths of the ice-covered oceans of Jupiter's moon Europa? We may soon know, but while we are waiting to find out, we can anticipate how this might have happened by exploring similar environments on Earth.

Let us start with the familiar world around us and proceed step by step to ever-more-alien environments in which conditions become less and less likely to support life as we know it. At the end of our journey, we will find ourselves in a hot, dark, stifling place that is virtually indistinguishable from the inferno itself. Along the way, the animals and plants that we know will quickly fall by the wayside, leaving only organisms that are chemically versatile enough to survive such hellish conditions.

Life in Unfamiliar Worlds

We will begin our journey in the open air and sunshine of White Sands National Monument in southern New Mexico. The monument was established in 1933 to protect the white sands themselves, an unusual and ephemeral desert environment that stretches for miles along the bottom of a shallow valley. Like everywhere else on the planet, humans have intruded, in this case violently. It was within a remote part of the surrounding region that had been set aside to protect the sands that the first atomic bomb was exploded.

The valley floor is covered with a huge expanse of brilliant, snow-white dunes. Around the periphery of this area, a variety of animals and plants have managed to invade the dunes, but hikers who venture further in soon leave them behind. The landscape they find there is an eye-aching and unrelieved white, broken only by the blue sky and the bare bones of desert mountains on the distant horizon. Sometimes the hiker will stumble on a cluster of cameras and film crew, in the middle of filming a low-budget Arctic adventure. The actors, broiling in their parkas, toil up and down the dunes. In the resulting film the dunes will be indistinguishable from snow, and audiences will be none the wiser.

The dunes at White Sands are not made up of the type of sand with which we are familiar. They consist of tiny crystals of gypsum or alabaster, which is simply hydrated calcium sulfate. Each of the crystals has been formed through evaporation in desert salt pans lying to the south and has then been picked up and blown here on the wind. The crystals may seem dry, but they feel sticky to the touch—they are continually sucking moisture from the desert air, and their surfaces are liquefying and crystallizing repeatedly. This process, along with the wind that is continually sculpting the great dunes, will soon erase the footsteps of the intrepid Arctic explorers and their film crew, leaving behind a blazing monochromatic stillness in which no bird sings.

And yet there is life in this unlikely place. Dig a few millimeters down into the gypsum, and you will find a thin, almost invisible, pale green layer. Close examination shows that the layer is made up of microorganisms that have found a safe haven from the drying wind and that can manage to multiply in the filtered sunlight that penetrates a little way into the surface layers of the dunes.

These cells live together in their own tiny ecosystem. The most important of them are several different types of blue-green algae, which are capable of making complex organic molecules through photosynthesis. Some of them are close relatives of the common single-celled *Synechococcus*, and others are related to the filamentous *Anabaena*. The desert sunlight provides enough energy for them to grow and multiply, enabling them to secrete surplus organic molecules into the salt-rich liquid surrounding the gypsum crystals. These nourishing molecules are quickly taken up by other members of this tiny community.

In spite of the abundance of sunlight, this layer of life beneath the dunes is a sparse one. This is because the wind-blown gypsum contains no nitrogen, an element absolutely essential for all life. Nitrogen is an important part of proteins and nucleic acids, and without it the algae in this frail layer could not exist.

Ironically, like sailors on a raft surrounded by undrinkable seawater, the algae are surrounded by plentiful nitrogen but they are unable to use it directly. Four-fifths of the atmosphere is gaseous nitrogen, but it must be converted to soluble ammonia and then preferably oxidized to nitrate before it can be used.

Unlike the sailors, the algae have evolved mechanisms for dealing with their problem. Accomplishing the conversion to ammonia is a difficult and energy-intensive process, and in this little ecosystem it is carried out chiefly in specialized thick-walled cells that stud some of the algal filaments. Inside these cells, sequestered safely away from oxygen, nitrogen-fixing enzymes pluck the nitrogen gas from the air and add hydrogen to form ammonia. We know that this ability evolved at least two billion years ago, for traces of such cells have been seen in microfossils dating from that time.

The algae are not the only nitrogen fixers in this tiny community. Another type of cell that lives along with them, the nonphotosynthetic bacterium *Azotobacter*, can make ammonia from the air without the need for such thick-walled cells. *Azotobacter* makes unusually oxygen-hungry enzymes that can keep the poisonous oxygen away from this essential process.

The algae and bacteria of this tiny world must use much of the energy provided by the Sun in order to make this tiny trickle of ammonia. They must use another substantial part of the Sun's energy to battle the tendency of the gypsum to suck moisture out of their cells. Yet another demand on their energy budgets is the need to scrounge traces of essential elements like phosphorus, sulfur, magnesium, and cobalt from the almost pure crystals that surround them. After all this, there is very little energy left over for growth and reproduction, which explains why this little green layer is so sparse as to be almost invisible.

Living on a substrate of pure gypsum is not easy. But although theirs is in many ways a desperate life, these little cells do have the advantage of having the whole place to themselves. The various species that make up this simple ecosystem have managed to occupy one of the most extreme environments on the surface of the Earth. This means that even though they must still compete among themselves for sunlight, nutrients, and trace elements, they have no other competitors.

The blue-green algae and bacteria that live in the dunes of White Sands are exposed to the Sun and the atmosphere. Although their environment is an extreme one, they are still a part of the world with which we are acquainted. But as we move away from our own world and into even more extreme environments, we will meet creatures that become less and less familiar.

Some of these unusual ecosystems have been discovered in other types of gypsum deposits. Remarkably, some organisms that live in them can actually use the gypsum itself.

Oat Burners and Gypsum Reducers

The Camargue, the delta of the Rhône River, is flat country largely unsuited to grazing or farming, with many salt pans and shallow, briny marshes. Large stretches of it have remained relatively untouched as a result. This empty land is France's closest approximation to the Wild West, the resemblance heightened by the bands of wild horses that roam the area. This did not escape the notice of French filmmakers, who made many low-budget cowboy movies in the Camargue early in the twentieth century. There seems to be something about such extreme environments that encourages movie-making!

As those Gallic cowpokes rode across the misty plains and marshes of the Camargue, Gauloises dangling from their lower lips, their horses' hooves would often crunch through hard salt pan. This disturbed the thriving microbial communities on the fringes of many of the shallow ponds.

In these communities, far more complex than those at White Sands, blue-green algae, purple photosynthetic bacteria, and sulfate-reducing bacteria can all be found living in a complex stack of layers under a crust of gypsum. There they form a miniature, self-contained world that has many similarities to our own—and many differences as well. The ecosystem can be more complex than the simple community we saw at White Sands because the gypsum here in the Camargue is mixed with organic material.

To understand this world, it helps to remember some of the chemistry from Chapter 2. Living organisms manufacture additional living material by exploiting the release of energy from reduced compounds. Some anaerobic organisms, although they are able to live in the absence of free oxygen, can gain only limited amounts of energy from these compounds. Aerobic organisms, like ourselves, can gain far more by using oxygen. In effect we burn these compounds, albeit under highly controlled conditions. When our bodies burn carbohydrates such as sugars in the presence of oxygen, the ultimate result is carbon dioxide (oxidized carbon) and water (which you can think of as oxidized hydrogen).

As we saw in Chapter 7, it is by tapping into and controlling this process at every step that organisms such as ourselves gain energy. Some anaerobic organisms manage to gain at least some of this extra energy by using oxidizing chemicals available to them even in the absence of oxygen.

Of course, this cannot continue indefinitely, for the supply of reduced compounds will quickly become exhausted. In the world with which we are familiar, plants use photosynthesis to close the circle, employing the energy of the Sun to reduce or "un-oxidize" carbon dioxide and water. The result is a flood of energy-rich compounds such as sugars. Oxygen is also produced, and it is used by both plants and animals to oxidize these newly synthesized compounds once again to gain energy from them.

At least, this is how the organisms with which we are familiar use the light of the Sun to close the loop. But light does not penetrate everywhere that life can survive. There are other ways in which oxidized compounds can be reduced again, allowing thriving ecosystems to be maintained even in the absence of light.

The Sun is still the ultimate source of power for the little communities of algae and bacteria of the Camargue. Just as at White Sands, green algae live underneath a thin crust on the surface of the gypsum, where air and light can still penetrate. They carry out photosynthesis in the ordinary way. And they produce a plentiful supply of simple organic molecules that slowly diffuse down into the deeper layers of the crust, nourishing other organisms that live there.

Beneath the algae, where oxygen from the atmosphere does not penetrate but there is still a modest amount of light, the next layer is occupied by bacteria that are able to carry out a second type of photosynthesis. Such photosynthetic bacteria use compounds other than water, prominent among them the foul-smelling hydrogen sulfide gas so often associated with this type of anaerobic life. The hydrogen sulfide is converted to elemental sulfur, a less energy-demanding reaction than the splitting of water. This means that these bacteria can still survive at low light levels and can wring power from low-energy light. Indeed, some bacteria of this type can live deep in lakes where little light penetrates.

Such bacteria do not need air, and some of them are actually poisoned by oxygen, but they still use light and the carbon from carbon dioxide to make energy-rich compounds. The oxygen from the carbon dioxide oxidizes sulfur into sulfate. The sulfate is excreted as a waste product. And sulfate, you remember, is a principal component of gypsum.

The hydrogen sulfide that these bacteria require has diffused up from the darker layers below them, in exchange for the organic molecules produced by photosynthesis that drizzle down from above. And it is here that the circle of life in this little universe is closed. The source of this hydrogen sulfide is sulfate-reducing bacteria, organisms that are anaerobic and nonphotosynthetic. They are the deepest-dwelling inhabitants of this little world and are approximate analogues of the humans and horses that live in the larger world above them. Unlike humans and horses, sulfate-reducing bacteria do not need oxygen, for they can utilize the great oxidizing power of gypsum to burn the food that has kindly been supplied by the photosynthesizers living above them.

Even though they are anaerobic, these bacteria can extract large amounts of energy from the life-giving organic molecules diffusing down to them. Indeed, they are as efficient at extracting energy from their food as air-breathing animals, because they can burn it completely by converting the oxygen rich sulfate of the gypsum to sulfide. And they pay their way by supplying sulfide to the bacterial photosynthesizers above, which otherwise would not be able to survive. This little community is wonderfully self-sufficient, although it does ultimately depend for its energy on the sunlight that bathes the gypsum crust.

Beyond the Familiar

Casual destruction is often wreaked on such tiny ecosystems by the trudging tourists and pseudo-Arctic explorers who leave their footprints on the White Sands, or by the herds of wild horses that splash through the shallow pools of the Camargue. But the destruction is local and easily healed. These tenacious and resourceful communities of microorganisms are likely to outlast both humans and horses. And they are not alone in their abilities to flourish in unexpected places.

Over the last few decades, as unusual living communities have been discovered and understood, it has become increasingly apparent that we must broaden our definition of where life can exist. We have discovered that living organisms have managed to colonize very extreme environments, from the boiling hot springs of Yellowstone to the submarine volcanic vents that spew out superheated water at the bottom of the sea. Thriving microbial communities have been discovered living just below the marble surfaces of ancient Greek statues, surviving on the baking hot stones of the California desert, and clinging to the freezing rocks of windswept Antarctic valleys. Understanding how microbes can live under such extreme conditions has broadened our comprehension of life's infinite resourcefulness.

One striking feature of such extreme environments is that they tend to slow down the pace of life. The bacterium *E. coli* is popular among scientists because, among other things, it can complete its life cycle in twenty minutes when it is grown on a rich food source. A single cell of this profligate creature can produce billions of progeny overnight. But organisms living on the edge of the possible may take much longer to multiply and colonize their environment.

Lichens on tombstones may take decades to grow an inch in diameter. An even more striking example of the slowing of metabolism under extreme conditions is provided by the accumulation of desert "varnish." For many years, generations of geologists had assumed that the dark, polished layer often seen on the rocks of a stony desert was some-how due to weathering. Now, however, the varnish is known to be bac-terial and fungal in origin. These organisms, able to survive the baking heat and almost total absence of water, make the dark pigment by slowly oxidizing the manganese metal in the rocks to dark manganese oxide. Why they do so is still unclear, but they may accumulate these deposits as protection against the ultraviolet light of the pitiless desert sun.

The varnish on rocks in the west of Australia has been shown to have taken thousands of years to accumulate. In Southern California, early Spanish missionaries made huge crosses in the desert by turning the rocks over so that the varnish was no longer visible. After three hundred years, varnish has only just begun to appear on the rocks' upper sur-faces again. Such slow-growing organisms give us some inkling of an early world in which the first protobionts must have grown just as slowly, cadging what nourishment they could from an unfriendly environment.

Far from the Sun

All the tough organisms that we have met so far ultimately draw their energy from the Sun. Are there any that do not? Two remarkable dis-coveries suggest that such organisms may exist, but that true inde-pendence from the Sun is very difficult to find. In our search, let us plunge first into the depths of the oceans and then into the depths of the Earth.

The submersible *Alvin*'s 1977 discovery of previously unknown or-ganisms living in the dark around deep ocean hydrothermal vents gen-erated a huge research effort to understand how this unique ecosystem sustained itself. As discussed in Chapter 4, the vent oases were found to be teeming with tube worms, shrimps, crabs, mussels, and even fish, something completely unexpected.

Because these deep-sea ecosystems thrive in total darkness and gain their energy from the hydrogen sulfide and methane emerging from the rocks, some researchers have suggested that they are completely inde-pendent of the world far above them. But it turns out that they are not.

The large animals such as tube worms and crustaceans that live there are not on some separate branch of evolution. Instead, they are de-scended from animals that once lived in shallower, sun-filled waters and that colonized the depths relatively recently. Now, like cave-dwelling animals, they have become specialized inhabitants of a sunless world.

Their lives are not without risk. Sometimes, as happened recently off the coast of the state of Washington, an undersea vent can explode, with devastating effects on the vent communities. Sometimes the vents on which the bacteria and animals depend can simply cease to spew their nutrients. But although the communities themselves may die, bacterial spores and larval forms of the animals drift from vent to vent, colonizing new vents as old ones disappear. And sometimes the descendants of at least some of the animals can return to shallower waters. Clams that draw their energy from the bacteria living inside them have been found in shallow water polluted by sewage near Los Angeles.

The bacteria that live inside the tube worms and around the vent openings also have a strong connection to the surface. New evidence suggests that the worm–bacteria symbiosis may have been established originally in shallower waters and only then moved into the deep sea.

Unlike the photosynthetic bacteria in the gypsum communities of the Camargue, these bacteria cannot generate complex reduced compounds in the absence of free oxygen. They live in Stygian darkness and so cannot use light energy to split carbon dioxide. Instead, they require oxygen to oxidize the sulfide that comes from the vents, enabling them to generate energy-rich compounds by using the reducing power of the sulfide. This oxygen has in turn been produced by photosynthetic organisms in the light zone far above, where it diffuses into the depths. Some bacteria in these communities can use nitrate instead of oxygen, but even the nitrate has its origins in the nitrogen-fixing plankton of the world above.

Remote and alien as these communities are, they still have connections to the world we know. We must penetrate into even more unusual environments before we find organisms that have broken this connection—or that perhaps never had any connection to the surface in the first place.

Severing the Connection

One of the most startling and controversial recent findings in science has been the discovery that the rock far beneath our feet harbors enormous numbers of microbes. Thomas Gold of Cornell University has suggested that in sheer mass they equal and perhaps exceed the mass of organisms that live on the Earth's surface!

This discovery was a long time in coming. In the 1920s, several scientists claimed to have found living bacteria in lumps of coal taken from deep mines. These bacteria could not have been contaminants, they supposed, because the surfaces of the coal lumps had been carefully sterilized when they were brought into the laboratory. One of the early claimants, bacteriologist Charles Lipman, thought that bacterial spores had somehow managed to survive in the coal beds ever since they were laid down during the Carboniferous, more than 200 million years ago. (Recall from Chapter 1 that Lipman also claimed he had found bacterial spores in meteorites.)

It was soon shown, however, that whenever living bacteria could be cultured from lumps of coal, it was not because they were the original inhabitants, but rather because the coal seams had been fractured so that water could percolate down into them. Unfractured, uncontaminated coal really did turn out to be sterile. Lipman had not discovered bacteria that had somehow survived from before the age of dinosaurs. Instead, these bacteria had penetrated far more recently into the coal seams from the world above.

Although contamination turned out to be the source of these coal microbes, it was astonishing that the slow percolation of water from above had managed to introduce living bacteria into the coal seams at such great depths. Though most or all of the inhabitants of these bacterial communities certainly did originate at or near the surface, the communities themselves probably persisted for thousands of years and perhaps even longer.

Further, it was soon shown that these long-buried bacteria are not a random sample of the bacteria that live on the surface. Living far from the Sun, they have apparently established their own communities with their own rules. Like the adaptation of hydrothermal vent communities to the deep ocean, they have become adapted to life in the deep rocks.

How did they get there, and how do they survive? In the decades following these microbiological discoveries, geologists found evidence that the rocks deep beneath the surface are anything but quiescent. In every part of the continental crust, the rocks are constantly in motion, as a result of tectonic plate movement and subduction. The process is slow but continuous. It carries rock from the surface down to the mantle far below and back up again in cycles that can take up to a hundred million years.

Sometimes, particularly in volcanic or earthquake-prone regions, this slow cycle is short-circuited. In these places, rock strata can be suddenly buried by lava flows or fault movements, only to be thrust to the surface again or uncovered by erosion after periods ranging from a few hundred to a few million years. We saw earlier how such events can have profound effects, even acting as triggers for ice ages.

The rocks that are caught up in these cycles are often fractured by earthquakes. These fractures allow fresh supplies of water to be introduced from above. Any living organisms that the water contains will be supplied with energy-rich materials from the newly fractured rock. These materials are of two types. Some, like coal, were originally derived from living organisms. Others, in the form of gases such as hydrogen, hydrogen sulfide, and methane, diffuse upward from the Earth's interior. This second group of materials is geological rather than biological in origin. Thomas Gold has suggested that this upward flow is much more plentiful than geologists had previously supposed and that it comes from huge deposits of oily hydrocarbons that accumulated during the formation of the early Earth.

Because heat increases with depth, there is only a relatively narrow zone in which life can take advantage of these energy sources. As we saw in Chapter 3, heat is generated in the Earth's interior mainly by natural decay of radioactive elements and to a lesser extent by tidal friction. This heat is considerable: The temperature at the Earth's core is estimated to be between 4,000° and 4,500°C, only slightly cooler than the surface of the Sun.

Life as we know it probably cannot exist at temperatures much above 150°C, which means that it must be confined to a thin skin on the planet's surface. Even a thin skin, of course, is a gigantic volume of rock. Gold is one of the first to call attention to the huge evolutionary potential of communities of organisms that live far underground. He has even suggested that the deeper, hotter layers may harbor life as we *don't* know it. Perhaps, far beneath our feet, simple organisms can somehow thrive in the terrible heat. Because they would be so different from the creatures that we know, Gold suggests that we have not yet been clever enough to find them. Although this speculation has been greeted with considerable skepticism, we should remember before dismissing it out of hand that the thriving bacterial communities that survive at temperatures near boiling were ignored or unknown just a few decades ago.

Coal mines are not the only source of deep-rock organisms. The deepest bacterial community yet discovered was found in 1994, the result of a deep drilling project actually inspired by Gold himself. He had championed the idea that oil and natural gas originate far beneath the

Earth's surface and were not formed by living organisms. This was an idea first put forward in the nineteenth century by Dmitri Mendeleev (see Chapter 2). As these materials percolate upward, living organisms in the deep rock use them for energy and leave their chemical imprint. This fools us into thinking that the origin of coal, oil, and gas has been entirely biological.

Gold thought that an ancient meteorite crater in central Sweden, called the Siljan Ring, would be an ideal place to look for gas and oil. The ring is the result of an impact that took place 380 million years ago. The meteorite cracked the Earth's crust, and even after all this time the granite rock still shows numerous fractures. In the center of the ring there is no sedimentary rock, so if oil or gas are to be found there they must presumably be of nonbiological origin.

In a remarkable display of persuasiveness, Gold inspired the Swedish government to subsidize two boreholes in this otherwise unpromising region. In 1988, when the boreholes reached almost 7,000 meters—as far below the surface as a substantial Himalayan peak rises above it—some very odd material was brought up.

This material was a thick, sludgy, oil-like substance filled with tiny particles of magnetized iron. Its most immediately obvious property was that it stank—it reminded Gold of a long-dead rat. Chemical analysis showed that it had both a nonbiological and a biological origin. It was intriguing that the tiny particles of magnetite that it contained resembled similar particles that are manufactured by many present-day bacteria to orient themselves to the Earth's magnetic field.

As the drillers battled this sludge, which kept clogging up the drill pipe, they did manage to bring up eighty-four barrels of what appeared to be perfectly ordinary oil. Oil from so great a depth and from so unpromising an area geologically would seem to reinforce Gold's idea that most such deposits are nonbiological and come from deep within the Earth.

Alas, gas was not recovered in appreciable quantities, and according to a 1999 account by Gold, almost nobody believed that any oil could possibly have been found, and indeed much of it was probably drilling lubricant. Gold's theory, already regarded with amusement and disbelief by most geologists, suffered a setback from which it has not yet recovered.

Nonetheless, the boreholes yielded some large scientific payoffs. Although most of the lower part of the hole was quite dry and showed no clear indication of living organisms, the middle regions turned out to support life. The microbiologist Karsten Pedersen (see the introduction) and his colleagues at the University of Goteborg discovered a thriving bacterial community at 3,500 meters. At this depth the granite was still fractured, so that water could penetrate. The temperature of the water was a sizzling 70° C, hot enough to scald.

These bacterial communities were so far below ground that their connection with living communities on the surface was tenuous at best. Yet the bacteria were surprisingly abundant, especially when one considers that the thousands of meters of fractured granite above them should surely have filtered out any organic material from the surface. After all this filtration, the water that reached this depth should have been as clear as vodka. Surprisingly, it turned out instead to be yellowish and a brisk generator of hydrogen sulfide.

Where did these organisms get their food? A few bacteria that the Swedish group found were able to grow in the laboratory on simple organic molecules, so it was conceivable that at least some of these nutrients had worked their way down from above. There may have been at least some connection to the surface.

A year later, Todd Stevens and James McKinley of the State of Washington's Pacific Northwest Laboratory found an alternative answer to this question. Their study of the deep rocks in south central Washington was made possible because of growing concern about the Hanford nuclear plant located in the area. The plant is surrounded by festering dumps of radioactive wastes, some of it leaking from rusting million-gallon steel containers. Boreholes have been drilled throughout the region to find out whether these wastes are being carried under the ground to surrounding communities and to the nearby Columbia River, and whether deep aquifers might have become contaminated.

This part of Washington, near the Cascade chain of volcanoes, has been volcanically active for a long time. When boreholes are drilled there, they penetrate layer after layer of old lava flows, passing through a series of aquifers that have accumulated between the flows like the fillings in a sandwich. Even though some of these aquifers have been isolated from the surface for at least 35,000 years, Stevens and McKinley found thriving microbial communities in them.

Contamination from the drilling itself could not be ruled out completely, but the microbes that they isolated had the properties to be expected of the inhabitants of deep-rock communities. They thrived at high temperatures, and the bacteria from a given depth grew best on the energy-rich compounds that were also found at that depth.

The simplest and most pervasive of these energy-rich compounds was hydrogen gas. When Stevens and McKinley crushed samples of lava from the area and mixed the resulting powder with water, it produced gaseous hydrogen. The hydrogen was not biological in origin, but came from the lava rocks themselves. The researchers realized that the rocks of the whole region formed a vast chemical generator of hydrogen gas.

When they mixed sterile, crushed samples of this rock with water from the deep boreholes, they found that some of the bacteria from the water thrived. These bacteria, able to use hydrogen itself as an energy source, greatly outnumbered and tended to outgrow those that required more complicated compounds to survive.

This work has sparked controversy, but if it holds up, it will be strong evidence that living organisms can exist completely cut off from the surface. Hydrogen-utilizing bacteria are unlikely organisms indeed, with capabilities that astonish us. Some of them can grow on nothing but rock and water, along with hydrogen, carbon dioxide, and nitrogen gases, a diet of such spartan simplicity that it would put any vegan to shame.

So far, only a few bacteria have been found that can survive under these ascetic conditions, and they grow very slowly, since it is very difficult for them to make the carbohydrates and other molecules needed for life. Nonetheless, they do grow. Pedersen's group has managed to culture some of these bacteria from granite taken from the Swedish tunnel that we explored at the beginning of the book.

In spite of these exciting discoveries, we know very little about these deep-rock communities. The simplest questions about them have yet to be answered. For example, are these bacteria rare or are they numerous?

When we consider the planet as a whole, the deep-rock communities are likely to be plentiful. This is because these dramatic discoveries have been made in deep rocks in both the United States and Sweden, from sites that contain rocks of very different types with very different origins. And many other microbial communities are being discovered in deep rock sites around the world, although most of the cultured microbes that have been found in them seem to have at least a tenuous connection with the world of the surface.

Certainly, the volume of the Earth's crust in which these bacteria might exist is huge. Most of the crust is of course solid rock, but it contains enough water-filled cracks and fractures to support uncounted numbers of microorganisms. To find out how many such cracks might exist, Gold computed the volume of rock in the crust that is cool enough to support life. He estimated that about 3 percent of this volume is water-filled pores and cracks where living organisms could survive, and guessed that bacteria might occupy 1 percent of this space in turn.

The result of his calculation is that if all the deep-rock bacteria living in this immense volume could be collected and spread out on the Earth's land surface, they would form a yucky layer about 1.5 meters thick— certainly at least comparable to the volume of the living organisms that we see around us. Even if Gold has overestimated the quantities of these microbes by a factor of ten or even a hundred, there must still be immense numbers of bacteria in that hidden, subterranean realm. The possibilities for evolution in the depths of the Earth are enormous.

A Byronic Catastrophe

In our journey to the center of the Earth, we have reached one of the goals that we set out at the beginning of this chapter. Life is indeed possible under conditions wildly different from the world we know. Life exists at depths where high temperature, darkness, and the absence of the necessary nutrients for surface life would seem to make life impossible.

Hydrogen-eaters are the very kinds of bacteria that we would expect to survive in a place that is utterly cut off from living organisms anywhere else. But are they really independent of life on the surface? Neither the American nor the Swedish groups could rule out the possibility that a tiny amount of organic matter had managed to penetrate into these deep rocks, thus connecting them with the world above.

If there really is no such link, then the organisms in the deepest parts of the boreholes must be generating their energy entirely from gases supplied by the Earth. They would be totally independent of energy from the Sun. Stevens and McKinley came up with a wonderfully evocative acronym to designate such a community: a SLiME, which stands for Subsurface Lithoautotrophic Microbial Ecosystem.

The test of a true SLiME is what might happen to it if the Sun were to be extinguished. Suppose that some great catastrophe, of the type that Byron wrote about in his poem *Darkness*, were to plunge the Earth into perpetual night. Although Byron imagined that the Sun could be snuffed out like a candle, we know today that this would be impossible. But it is less far-fetched to suppose that the Earth could be perturbed by a near-miss encounter with a giant rogue comet and be sent hurtling away from the Solar System, like the new planet that has recently been spotted being flung out of a nascent solar system in Taurus (Figure 3.2).

The interstellar darkness and cold would quickly kill all the organisms on the Earth's surface. Free oxygen would eventually disappear, and a thick layer of ice would form on the oceans. They would not freeze completely from top to bottom, because the internal heat of the Earth would keep deep hydrothermal vents functioning. Near these vents, liquid pockets of water would remain under the ice. But, in the absence of a fresh supply of oxygen, it seems likely that all the organisms living on the ocean bottom would eventually disappear as well.

In the face of such a Byronic catastrophe, very few living organisms would have a chance of surviving. The most important of these would be bacteria that have incredibly simple nutritional requirements. Some of these might be the hydrogen-eating bacteria.

A well-studied candidate for such a versatile organism is *Methanobacterium thermoautotrophicum*, which is among the most self-sufficient creatures ever found. This rod-shaped microbe was first discovered in 1971, living in sewage sludge, where, like the deep-rock organisms, it can thrive at temperatures as high as 70°C. The bacterium can manufacture everything it needs from three simple gases: hydrogen, carbon dioxide, and nitrogen. Along with water and various elements such as calcium, iron, and sulfur that it can glean from the surrounding sludge, these gases are sufficient. It has no need for free oxygen.

The usual venue of this self-reliant organism is certainly not sewage sludge. Although the microorganism may not be part of a SLiME, some of its equally resourceful relatives probably are. All the components that these relatives of *Methanobacterium* need to survive and multiply are available in the deep rock and have been available since the formation of the Earth.

After a Byronic catastrophe, SLiMEs would presumably survive for as long as the Earth generated enough internal heat to keep some of the water trapped in the rocks liquid, and for as long as the rocks themselves were capable of generating precious hydrogen gas and carbon dioxide. Even though the hydrogen-eaters could only metabolize with painful slowness, they might still support a few hangers-on in the form of heterotrophic bacteria that could dine on the meager leavings from the hydrogen-eaters' table.

SLiMEs might survive for enormous periods if, as seems likely, radioactivity-generated heat from the Earth's core was able to keep the convective flow in the mantle going, bringing fresh supplies of hydrogen and other gases near the surface. The Earth could drift for billions of years through interstellar space, carrying these bacterial communities with it, until the final exhaustion of its radioactive elements. Life would die, not with a bang, but with an exceedingly long, drawn-out whimper.

This ability of a SLiME to persist indefinitely in the depths of the Earth does not mean, however, that life itself necessarily began in SLiMEs. *Methanobacterium* lives on the surface of the Earth, and all the bacteria identified from the deep rocks have relatives on the surface. Indeed, we have been able to penetrate further toward the base of the tree of life by examining bacteria that live on or near the surface than when we examine bacteria from the depths of the planet.

Tracing the Bacterial Tree to Its Roots

Norman Pace, now at the University of Colorado, and his many collaborators have helped to trace the evolutionary tree of living organisms back to its roots. They have done this by taking samples of water and scrapings from the rocks of hot, steaming pools in Yellowstone Park and other volcanically active areas. The samples must be obtained with great care, because the pools release quantities of poisonous hydrogen sulfide and carbon monoxide—in 1979 an entire village in Java was wiped out by such an invisible cloud. Incautious investigators can also be parboiled, as happened to Cyril Ponnamperuma when the apparently solid crust near the Icelandic pool he was investigating collapsed suddenly.

The microbial communities that Pace and his colleagues found in these pools are not as exotic as SLiMEs, but they give a glimpse of a world at least as old. Studies of DNA show that some bacteria in the hot pools have very ancient evolutionary roots that can be traced back to near the base of the tree of life as we know it (Figure 8.1).

The discovery of ancient life-forms in these places does not mean that life originated deep in the rocks or in hot springs. We must remember that primitive as these bacteria seem, they have gone through billions of years of evolution. Their ancestors may have migrated many times back and forth between the surface and the deep rocks. There is no way of telling where their ancestors first appeared or how many things have happened to them during this long evolutionary history.

Further, although some of these bacteria do lie near the currently hypothesized base of the tree of life, this base is simply the deepest ancestor of the organisms that we presently know about. There may be

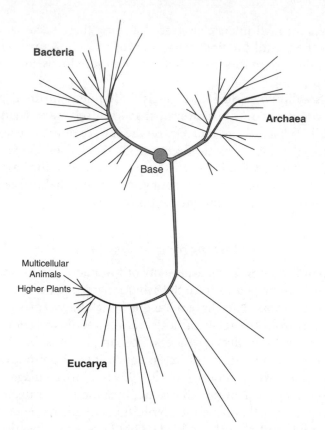

FIGURE 8.1   The tree of life. We are looking down
on the tree from the top, and the base of the tree is in
the middle. You can see that higher animals and
higher plants form only a tiny part of the tree.

another and deeper base to the tree, giving rise to major branches that
lead to vast and wholly undiscovered groups of organisms. And we have
no idea where or even whether we will find such organisms.

All these problems may be temporary. Our understanding of the tree
of life has expanded enormously in the last few years, and this new
knowledge is now being applied to many different ecosystems. One
large difficulty, which the new techniques are now just beginning to
overcome, is that the bacteria isolated from deep rocks and hot springs
are distressingly coy about being studied. Only a tiny fraction of these
bacteria, less than 1 percent, can be grown in the laboratory—the rest
seem unable to emerge from a state of dormancy, or die when trans-

planted to petri dishes, in spite of the best efforts of microbiologists who want to study them.

This problem is not unique to unusual bacterial communities. Generations of microbe hunters have confronted it whenever they have tried to study natural microbial ecosystems, whether the organisms be from ordinary soil, mud, thermal pools, or even the guts of humans and other animals. Only a small fraction of the microbes that live in these thriving mini-ecosystems can be persuaded to grow once they have been isolated in the laboratory.

This means that we know very little about how much diversity is harbored by even the best-studied microbial ecosystems. The extent of our ignorance was vividly brought home recently when the polymerase chain-reaction (PCR) technique was first used in an attempt to identify at least some of the many different species of bacteria that stubbornly refuse to grow.

PCR is an ingenious biochemical trick (Figure 8.2) that allows scientists to amplify a few copies of a stretch of DNA into many copies, yielding enough to determine their sequence. These days, quite long pieces of DNA can be sequenced in this way. The trick requires, however, that at least some small portions of the DNA sequence that is being searched for are already known. This is because the PCR technique requires the manufacture of short pieces of DNA called "primers" that match known parts of the otherwise unknown DNA. These primers can then be used to get the process going.

PCR can be done with any sample of DNA, including samples extracted directly from soil and rock. Because these samples contain bits of DNA from organisms that obstinately refuse to grow in the laboratory, PCR will at least force them to reveal some of their secrets.

One gene has been particularly useful in probing the branches of the tree of life. This is the stretch of DNA that codes for the 16S ribosomal RNA, which specifies a part of the protein-manufacturing ribosome. (16S is a measure of the size of the molecule.) The gene's function is so important that parts of it have evolved very slowly, leaving short stretches of it almost unchanged. Very distantly related organisms that have been separated by long periods of evolutionary time still carry virtually identical DNA sequences in these short regions, even though the DNA in other less conserved parts of the gene has undergone many changes. These short, conserved stretches make ideal primers, because the same primers can be used to amplify the genes from a wide variety of organisms.

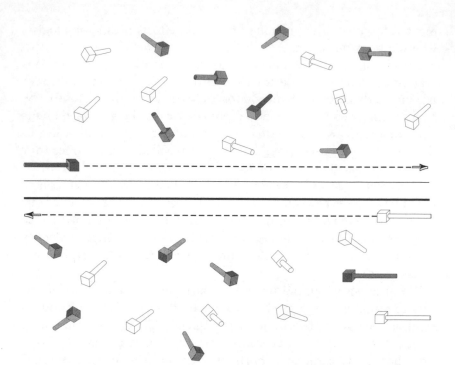

FIGURE 8.2   In the PCR reaction, a piece of DNA (shown stretched out in the middle of the picture) is mixed with large numbers of small bits of DNA called primers, which complement the sequence at the ends of the piece. The double helix of DNA is then separated into its two halves by heating it. Afterward, as the temperature drops, complementary primers attach themselves to each of the halves. An enzyme can then use these primers to trigger the synthesis of new complete complementary strands (dotted lines). After this, the DNA strands are separated again and the cycle is repeated. With each repetition of the cycle, the number of copies of the original piece of DNA grows exponentially.

Not all organisms share these stretches, however. If the ribosomal RNA gene of some unknown organism has diverged so far from the sequences of known genes that even its most conserved stretches of DNA have changed, then PCR cannot be used to amplify the gene. The PCR primers will have nothing to match to, and the PCR reaction will not work. Such very different organisms will remain for the moment beyond our ken.

Some of the most successful searches using PCR have been carried out in some of the places we have already visited: the hot sulfur springs in Yellowstone National Park, and the deep boreholes in the Swedish granite. Other success stories have emerged from the mud of river estuaries around the world. During each search, the unexpected has happened.

The first unanticipated finding was that various groups of organisms were living in places where they should not have been found. One of these groups is the archaea, which form their own distinct branch of the tree of life (Figure 8.1).

Archaea had been discovered decades earlier by Carl Woese of the University of Illinois. Most strains of archaea that he originally looked at had been isolated from extreme environments that were very salty or acid or hot—or all three! They had originally been supposed to be merely unusual bacteria, but Woese found that their genes are very different from those of ordinary bacteria. Because of this, and because they lived in such unusual places, he understandably thought that they were direct descendants of the very earliest forms of life. He called them archaebacteria.

Alas, these archaebacteria turned out not to be quite as different from other groups of organisms as Woese had thought. Although, as we will see, they do carry very old genes, they may not be descendants of the very oldest organisms. Rather they may be a kind of pastiche of genes from various creatures, some very old and some much younger. Because they have turned out not to be the ancestors of the bacteria, they have now been renamed archaea.

When mud from river estuaries was examined using the PCR technique, the result was a big surprise. The mud came from a rather ordinary environment, where archaea should not live, and indeed archaea have never been cultured successfully from such samples. Nonetheless, traces of genes belonging to the archaea were found.

This discovery shows that the archaea are far more ubiquitous than had been supposed. The reason that they cannot be cultured from estuarine mud remains a mystery. Perhaps they are employing some very unusual compound for energy, or perhaps they are actually living inside other organisms. The PCR technique, ingenious though it is, cannot distinguish among such possibilities.

In the dark and very hot waters of the Obsidian Pool of Yellowstone Park, Pace and his collaborators have found the reverse situation. One might suppose, because this environment is an extreme one, that most organisms living there would be archaea. But in fact it was found that the bulk of the organisms living there consisted of many different kinds of true bacteria, some of which formed entirely new branches of the bacterial part of the tree of life. In a single study, Pace's team almost doubled the number of known groups of these true bacteria, adding twenty major new branches.

Karl Pedersen and his co-workers found a similar pattern among the bacterial communities that live almost five hundred meters down in the Swedish granite. When they examined these deep-rock communities by PCR, they too found new branches on the tree, branches that led to different and previously undescribed organisms. Some of these were archaea, and some were true bacteria.

Do these new discoveries mean that the full extent of the diversity of these ecosystems has been thoroughly explored? This is unlikely, for three reasons. First, as we saw, PCR only works with genes that have at least some short DNA sequences in common with the same genes from more familiar organisms. Bacteria that are even more alien than the ones found by Pace and Pedersen would be missed in a PCR survey. Because of this limitation on the PCR technique, it is certain that many more branches of the tree of life remain to be discovered.

Second, there may be living organisms in these communities that do not have the 16S ribosomal gene at all and that might even make their proteins by some completely different mechanism. If such creatures refuse to be cultured, they will continue to be—at least for the time being—undiscovered and unexamined.

Third, even though the genes of these new bacterial species have been found, we know little else about them because most of them obstinately refuse to survive in the laboratory. They turn up their noses at everything we offer them to eat. Perhaps they require compounds that we have not yet tried, or perhaps they can only grow in the presence of other organisms, in symbiotic relationships that we do not yet understand.

If some of those organisms are very different from anything we know, this might give us a glimpse of the kind of complex interactions between species that must have taken place early in the evolution of life. We will explore some of these interactions in the next chapter.

New Possibilities for the Origin of Life

If we are to find living organisms somewhere on our planet that retain extremely primitive characteristics, the most likely place to look for them is in unusual microbial communities. The race is on to find the most primitive, or at least the most biochemically different, living organism. But we must remember that this most primitive of organisms may be found, not in the deep rocks, but on the surface, and that even if it is found in the deep rocks, it may not have originated there.

In the deep rocks few opportunities would have existed for an early proto-Darwinian sorting out of molecules into different parts of the environment, since there are no currents, waves, or tidal action. There would have been no light, which as we have seen is a very promising source of energy for the beginning of macromolecular synthesis. And there would have been no abundant, ever-renewing source of organic materials, like those provided by Miller-Urey reactions taking place in tidal pools and in the ocean. If protobionts evolved in the deep rocks, they would have had to depend on relatively puny amounts of raw materials that managed to filter down from above.

Taken all in all, the bacterial communities living in the deep rocks may turn out merely to be a fascinating example of the way in which life can adapt to extreme environments. The real origin of life may have taken place in a region that was not quite so alien. On our journey from the familiar to the strange, in search of the origin of life, we may be coming around full circle to the more familiar.

EVOLUTION BY COMMITTEE

My earliest complete statement of "serial endosymbiosis theory" was published after fifteen or so assorted rejections and losses of an early, painfully convoluted, and poorly written manuscript. . . . Today I am amazed to see a watered down version of [it] taught as revealed truth in high school and college texts.

Lynn Margulis, Symbiotic Planet, *1998*

LYNN MARGULIS, NOW AT THE UNIVERSITY OF MASSACHU-setts, has been a contrarian since her graduate student days in Berkeley in the early 1960s. She was distressed that geneticists knew nothing of the exciting new world of molecular biology, and that molecular biologists knew nothing of evolution. She wanted to try to link these very different ideas into a new synthesis, and succeeded beyond what were probably her wildest dreams.

What eventually resulted from her graduate work was a thorough marshaling of the arguments in favor of an old and seemingly discredited idea. Mitochondria, the powerhouses of our cells, are actually alien invaders. Our remote ancestors quickly learned to live with these invaders and to benefit from them. And this mitochondrial symbiosis is only one of many that have occurred in the course of evolution. When her idea was, after many vicissitudes, eventually published in 1970 by Yale University Press, its arguments were so compelling that the book catapulted the idea of symbiosis—organisms of different species living together—into the forefront of evolutionary theory.

Instances of bacteria invading living cells and apparently living in harmony with them had actually been found around the turn of the twentieth century by Russian botanists, particularly I. E. Wallin, A. S. Famintsyn, and K. S. Merezhkovsky. These observations of cooperation at the cellular level accorded well with the currents of Russian intellectual thought at the time—notably the philosophy of the revolutionary anarchist Peter Kropotkin, who had hoped that people could learn to cooperate without the assistance of government. This may explain why the idea of cellular cooperation was accepted in Russia and later in the Soviet Union, but largely ignored in the West. By the present time, however, the old ideas of the Russian botanists that Margulis has resurrected have turned out to be in their essence correct and they are currently accepted by capitalists and socialists alike.

In the decades since the appearance of her book, the possibility that symbiosis played an important role in the origin of life has been explored enthusiastically by many other scientists. Carl Woese, the discoverer of archaea, has taken this idea to an extreme. He thinks that life may have begun as a series of ever-changing, swapping committees of organisms or protoorganisms. These committees argued and jostled among themselves but eventually reached a kind of mutual consensus. In the process, they exchanged much genetic information. In such a world, it would have been hard to tell where one organism stopped and another began.

To understand how such strange ideas have gained credibility among some scientists, we must reexamine the tree of life. The tree that we encountered in the last chapter is less universal than we might think. When a tree is constructed using the 16S ribosomal RNA genes from a wide variety of species, it does indeed look like Figure 8.1. But the tree can look very different if we use different genes from those same species. Depending on the gene used, the tree's three major branches can emerge from its base in very different orders, and the branching orders of the twigs at the tips of the branches can vary greatly as well.

This means that as we move back in time and try to glimpse the origin of life, that origin becomes puzzlingly difficult to discern. There seems to be no overall clear-cut line of descent, no single tree that applies to all genes. It appears that different genes have had very different histories, in disagreement with the conventional Darwinian view of the world but in agreement with Woese's idea that organisms must often have exchanged genes at the dawn of life.

How can this have happened? Did the rules of genetics somehow not apply at the beginning of life? To examine the genetic circumstances under which life began and underwent its early history, we must bring Margulis's powerful ideas about symbiosis to bear, along with new ideas that are emerging about how evolution actually takes place.

The Early Days of Evolution

For many decades, the central assumption of the theory of evolution has been that it took place through what Darwin described as descent with modification. By this he meant that as one generation gave rise to the next through sexual reproduction, species became modified to become better adapted to their environments. Differences between organisms were sorted out by natural selection, with the fittest organisms always surviving and reproducing. The geneticists and evolutionary biologists who followed Darwin discovered that the modifications he postulated were actually caused by occasional alterations in genes, which appeared through the process of random mutation. It was generally agreed that if this process were to be continued over many millions of years, it would eventually give rise to all the bewildering diversity of life that we see around us.

Many people have found it difficult to understand how such a slow process could have resulted in the complex organisms and remarkable adaptations that we now see. This complexity is apparent at every level. We humans are an extreme example. We are highly differentiated multicellular organisms made up of at least 210 distinct types of cells. But even single-celled animals and plants can be remarkably complex as well, although their complexity has been achieved through a process of miniaturization rather than of cellular differentiation.

The single-celled ciliate *Paramecium*, for example, squeezes a sensory network, organs for mobility, a gullet, defensive structures, and various digestive and excretory structures into a little ovoid cell just barely visible to the naked eye (Figure 9.1). Some of its relatives even have a light-sensitive eyespot as well. These one-celled animals have managed to acquire most of the characteristics possessed by far more elaborate multicellular creatures such as ourselves.

Paramecium and the cells that make up our bodies have an important feature in common. Both they and the cells of our bodies are *eukaryotic*, which means that they have a true cell nucleus. This structure is usually located in the middle of the cell, where it sequesters the DNA-carrying chromosomes away from the hurly-burly of metabolism and protein synthesis that goes on in the surrounding region known as the cytoplasm. *Prokaryotic* organisms such as bacteria, on the other hand, do not have such a structure to protect their chromosomes. This complex internal organization of eukaryotic cells must have something to do with why eukaryotes have been able to become so complicated,

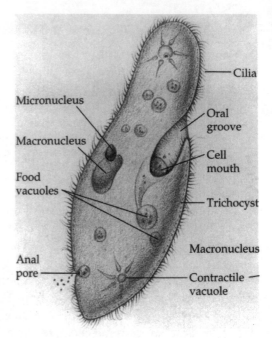

Cilia

Micronucleus

Oral groove

Macronucleus

Cell mouth

Food vacuoles

Trichocyst

Macronucleus

Anal pore

Contractile vacuole

FIGURE 9.1 *Paramecium*, a single-celled animal with many different structures that have equivalents in multicellular organisms such as ourselves. (Courtesy Neil Campbell)

though the connection continues to escape us. Even though eukaryotic cells have been on the planet for two billion years, approximately half the entire history of life on Earth, it is difficult to see how creatures as different as ourselves and paramecia could have evolved by the agonizingly slow accretion of tiny, favorable mutations one at a time.

Now evolutionists are beginning to realize that this kind of change is not the whole story. We are exploring the histories of our own genes and those of many other organisms in greater detail. In the process, we are starting to uncover evidence that there have been many exceptions to the dull regularity of ordinary evolution. The further we trace genes back in time, the more it appears that early life underwent many sudden changes that led it toward increased complexity. Such sudden changes are rarer now, but they still happen occasionally. And those early sudden changes may have played an important role in the origin of life itself.

Although genetic changes of large effect form an important part of evolution, there is no doubt that natural selection has been the prime motive force driving their acceptance or rejection. Whether the mutations are large or small, Darwin still reigns supreme. Our new under-

THE SPARK OF LIFE

standing of the evolutionary process has come, not through some over-throw of Darwin's ideas, but as a result of three major insights into how genes and organisms work.

Backbones and Insect Wings

The first of these insights is that mutations with large effects can fuel sudden evolutionary change. As a result of such substantial mutations, something as fundamental as a body plan can become drastically reorganized, leading to the appearance of new functions.

Such wild evolutionary "experiments" appear occasionally even today. Examples include the short-legged Ancon sheep breed, a mutant strain that enjoyed a brief popularity because of its inability to jump over fences. Ancon sheep soon fell out of favor because they had intractable health problems.

Other extreme mutants are the dramatically altered fruit flies that occasionally appear spontaneously in geneticists' laboratories (Figure 9.2). These mutants, which also tend to have poor survival, have taught us a great deal about how flies (and people) grow and develop.

Unless geneticists preserve these aberrant organisms with great care, virtually all of them tend to be weeded out very quickly and quite mercilessly by selection. Such mutants were dubbed "hopeful monsters" by the geneticist Richard Goldschmidt, who supposed that a tiny fraction of them might be lucky enough to have some characteristics so advantageous that they could overcome their disabilities and be able to survive and multiply.

Hopeful monsters that survive and change the course of evolution must be very rare today. But perhaps the likelihood of survival of such a dramatic mutant was greater during the early stages of multicellular life. In that distant time, the world was surely a simpler and less demanding place, in which the occasional hopeful monster could have found a niche.

Sometimes we can catch glimpses of possible hopeful monster mutations in the fossil record. Consider the plesiosaurs, oceangoing reptiles that lived during the age of dinosaurs and occupied an ecological niche similar to that filled by killer whales today. Like whales and dolphins, the plesiosaurs were the descendants of land-dwelling animals that had returned to the sea (Figure 9.3). Their adaptation to life in the ocean was marked by an enormous increase in the numbers of various body parts, particularly the vertebrae and the tiny bones that are the equivalent of

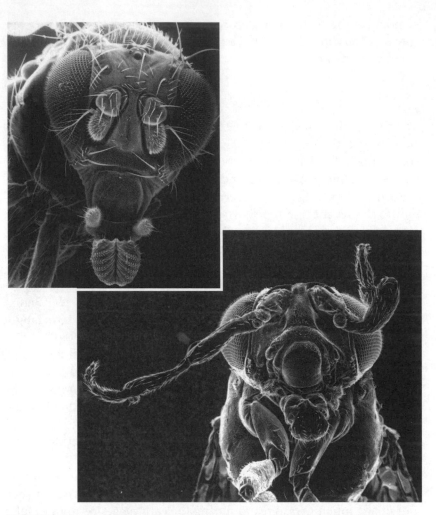

FIGURE 9.2 Scanning electron micrographs of the heads of a normal fruit fly (left) and one carrying the antennapedia mutation (right). In the mutant, the antennae have been replaced by legs. (Courtesy F. Rudolf Turner, Indiana University)

those in our fingers and toes. These long-necked, sinuous creatures, propelled by great flipperlike paddles made up of dozens of finger bones, could easily wriggle through the sea in pursuit of fish.

We now have strong clues to the kinds of gene mutations that might have led to such a huge multiplication of body parts. A gene has been found in mice that increases the number of vertebrae, and a mutant

FIGURE 9.3 This plesio saur, from the age of dino saurs, shows an enormous lengthening of the neck and tail, and a multiplication of the plesiosaur equivalent of finger bones. Such changes are unlikely to have happened one at a time.

gene that sometimes increases the number of fingers has recently been tracked down in humans. Of course, it is unlikely that the plesiosaurs suddenly acquired all of their extra vertebrae and finger bones by a single mutation of this kind—but it is also unlikely that they added all these bones painfully one at a time.

Developmental biologists are now catching other glimpses of genes that govern drastic differences in body plan. Sean Carroll and his colleagues at the University of Wisconsin have tracked down the genetic differences that determine whether insects have one, two, or three pairs of wings. The remote ancestors of all these insects appear to have had many more pairs. The oldest winged insect fossils that we know of, pterygotes from the upper Carboniferous some 300 million years ago, had large numbers of winglike flaps along most of the length of their bodies, which they apparently used for gliding.

Carroll has found that although changes in wing number were certainly extreme in their ultimate effects, they did not happen instantaneously. He has traced the reductions in wing number to several different mutations in the controlling genes. A variety of insect hopeful monsters must have appeared and been selected for during the course of these reductions, but probably none of them were too wildly different from their parents. If they had been, they would not have survived. Nonetheless, the cumulative effect of these substantial genetic changes was quite enough to bring about an acceleration in the pace of insect evolution.

Reticulate Evolution

The second source of sudden evolutionary change is one that chiefly affects prokaryotes, but that can also have dramatic impacts on our own lives. Many different species of bacteria can swap genes or entire clusters of genes, and such swaps can greatly increase their danger to humans.

Escherichia coli is normally a harmless or even useful denizen of our guts. But the infamous *E. coli* strain O157:H7, which has recently been responsible for a number of outbreaks of food poisoning around the world, has picked up sets of genes that help it invade the cells of the gut lining. Some of these genes are shared by virulent strains of *Shigella*, *Salmonella*, and other bacteria.

These borrowed genes are not the entire story of the heightened virulence possessed by this infamous strain, however. As its genome was investigated further, researchers found that it has borrowed an extra million bases of DNA from unknown sources and has pasted them into different parts of its chromosome. All this cutting and pasting seems to have happened fairly recently—strain O157:H7 is very much a work in progress.

Many species of bacteria can exchange genes that confer resistance to antibiotics. Sometimes, a number of these resistance genes are organized into tiny extra chromosomes called *plasmids,* which can easily be passed from one kind of bacterium to another. The plasmids themselves actually predate the introduction of antibiotics, for they have been found in strains of bacteria that were preserved early in the twentieth century. In those days, the plasmids were carrying other genes, some of which we know to be involved in the repair of DNA. We have no idea, however, what advantages (if any) those plasmids might have conferred on their recipients before antibiotics were discovered.

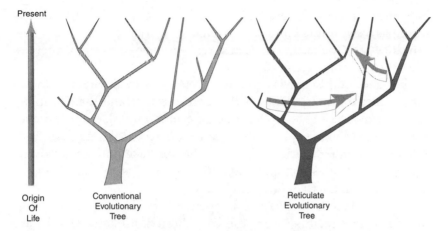

Present

Origin
Of
Life

Conventional
Evolutionary
Tree

Reticulate
Evolutionary
Tree

FIGURE 9.4   In a conventional evolutionary tree (left), some species go
extinct and some survive down to the present, but all continue to diverge
with time as the different lineages accumulate different mutations. But in a
reticulate tree (right), some of the genes on a branch can be transferred
between very divergent evolutionary lineages. In the tree shown, some genes
in lineages that have gone extinct have been "saved" by being transferred to
lineages that have survived down to the present.

One reason that this evolutionary fluidity was only discovered over the
last few decades is that it is not common in eukaryotic organisms like ourselves. In our small part of the overall evolutionary tree, there is very little
transfer of genes from one branch to another. Instead, the branches diverge in a regular way, and all the genes on the branches diverge together.

Among prokaryotes, although many genes also diverge over time in a
regular way, many others can be exchanged across branches. As the bacterial tree of life grew and the branches diverged, separate branches
sometimes fused whenever they happened to brush against each other
(Figure 9.4).

As you might imagine, this process of *reticulate evolution* can play
hob with the efforts of evolutionists to make sense of the tree of life.
But it can also give the evolution of prokaryotic organisms unparalleled
flexibility. Bacteria like *E. coli* need not painfully piece together genes
from here or there, but can pluck an entire set of advantageous genes
from other bacteria that share their environment.

There is no way of predicting when a set of genes that has evolved in
one bacterium may be inserted into another very different one, sometimes with disastrous consequences for our own species. This has

happened with the cholera bacillus. A set of genes that confers virulence has been inserted into its chromosome, probably as little as a few
centuries ago. One result has been seven worldwide cholera pandemics
since the end of the eighteenth century.

The genes in this arriviste piece of DNA are very different from the
other genes on the cholera bacillus chromosome, and their ultimate
source remains a mystery. This kind of evolution is able to hand over
ready-made genetic weapons, themselves the result of long periods of
evolution, to other species of bacteria that can take immediate advantage of them. It is as if muggers were to discover that submachine guns
are growing on the trees in Central Park!

Why does this not happen, or happen only rarely, in eukaryotes like
ourselves? Presumably we are too complicated. Prokaryotic cells are
simple cells. Even if they suddenly acquire a whole new collection of
genes, their developmental pathways are simple and robust enough to
accommodate the addition and even take advantage of it. Sudden and
massive genetic changes seem to be far more common in the prokaryotic world than the eukaryotic world, but they are not as obvious in
their effects because they do not result in deformed organisms. That
dangerous *E. coli* strain O157:H7 is a little rod-shaped cell, just like the
more familiar *E. coli* of the microbiology lab, even though the number
of its genes has been increased by 25 percent.

Chinese Boxes

The third source of sudden evolutionary changes is even more dramatic.

Natural selection is ultimately about competition and survival, but
sometimes the ability to compete and survive can increase as a result of
cooperation. Often, two species live together in intimate proximity for
mutual benefit, which makes them more effective competitors than if
they were surviving separately. The relationship is described by its
Greek term, *symbiosis*.

Symbioses are currently very common in the living world around us.
Lichens are a classic example. These rock-encrusting organisms are
made up of a combination of algae and fungi living together in intimate proximity.

This is a marriage of convenience, however, rather than a permanent
state of wedlock. The algal and fungal cells can clearly be distinguished
under a microscope, and most can be cultivated separately. The union
between the two kinds of organism has repeatedly been broken and re-

joined in nature as well. There is strong evidence that the algae and fungi readily come together when they are growing on a dry, exposed environment such as the surface of a rock, but in a richer and moister environment they have no trouble surviving separately. Exchanges of partners seem to have taken place again and again. Only a few species of algae are found as partners in lichen symbioses, but it is estimated that they are able to come to live together with over 40,000 species of fungi.

Other temporary associations are found throughout the living world. Some strains of the complex ciliated protist called *Paramecium* that we met earlier have engulfed tiny, single-celled green algae. Hundreds of algae can live happily inside a single *Paramecium*, photosynthesizing and providing themselves and their host with essential nutrients. When the paramecia are grown in the dark, they eventually lose their algae, but if light is reintroduced and they are presented with new algae, the symbiosis is readily reestablished. Although the paramecia prefer algae of the species that they originally carried, they can pick up other types of algae, sometimes from very different organisms such as green hydras, small multicellular animals that are also benefiting from similar symbioses.

Then there are more permanent relationships. Photosynthesizing algae that live inside the cells of coral polyps are largely responsible for the burgeoning exuberance of life on tropical reefs. Algae also live in the mantles of mollusks like the giant *Tridacna* and the heart cockle *Corculum*. The latter symbiosis is a very highly evolved one. The host cockles have acquired tiny, transparent windows in their shells, allowing beams of light to shine on the algae that live in their mantles.

When they are inside their hosts, these algae tend to be relatively featureless little green blobs, but when they are cultured separately, they soon revert to the appearance of their free-living ancestors. They grow flagella and other elaborate features and start to swim about.

Far more intimate and inseparable symbiotic relationships have been discovered, and two of them are absolutely essential to life as we know it. The best understood of these—the invasion of our cells by the bacterial ancestors of mitochondria—is also the best known.

How could such a symbiosis have begun? Some theorize that it perhaps began as a disease. Bacterial diseases that might lead to symbioses are found today. The cell biologist Kwang Jeon of the University of Tennessee has studied an accidental bacterial invasion of a laboratory strain of amoebas. The invasion took place in 1966, and at first the bacteria killed off almost all their hosts. After several years, however, the survivors were able to tolerate the bacteria very well. More remarkably, at the end of that time, the amoebas had somehow become dependent on the bacteria, for without them they could no longer undergo cell division. The bacteria make proteins that are exported to the nuclei of their host cells, and although the functions of these proteins are not yet understood, they appear to affect the hosts in important and fundamental ways.

In the space of a few years, the bacteria inside the amoebas have not only found a safe place to live, they have also managed to make themselves indispensable. They can no longer be expelled, for then their amoeba hosts would not survive. The evolution of this new symbiosis has taken place in Jeon's laboratory with astounding swiftness, a vivid illustration of how quickly symbioses can add complexity and new qualities to living organisms.

The discovery that our mitochondria had once been bacterial symbionts, along with the discovery that photosynthetic algae have repeatedly invaded both single-celled and multicellular animals, led to scientists' much more rapid acceptance of another hugely important symbiotic event that led to the appearance of green plants. The hosts of this ancient invasion were small, colorless, eukaryotic organisms. Like the cells of the fungal part of lichens, these eukaryotic cells began to live together with green bacterial cells that could supply them with plentiful energy from the Sun. It was not long before these little photosynthesizing bacteria, called blue-green algae or cyanobacteria, moved inside the cells of their hosts.

Not all the cyanobacteria invaded the cells of other organisms. You will still find their free-living relatives today, growing as dark green films on tree bark that has been shielded from the Sun or on the bottoms of damp flower pots. And they contribute significantly to the photosynthesis carried out by oceanic plankton, on which all the life on the planet ultimately depends.

These two symbioses, leading to mitochondria and chloroplasts, were enormously advantageous. The presence of mitochondria in both animals and plants gave a huge boost to the amount of energy that their hosts could extract from the environment. Chloroplasts extracted abundant energy from the Sun, at the same time pumping oxygen into the atmosphere and changing the world forever. Although chloroplasts and mitochondria are both examples of evolutionary cooperation, this does not mean that all of life consists ultimately of such warm, touchy-feely interactions (as James Lovelock had suggested in the first version of his Gaia hypothesis). When organisms acquired these new talents, they immediately became far more effective competitors, and far more

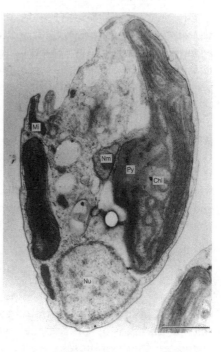

FIGURE 9.5 Electron micrograph picture of a slice through a cryptomonad, showing the red alga that has been engulfed by the cell. The red alga is the dark mass to the right of the cell. Nm is the alga's degenerate nucleus. The bar to the lower right indicates one thousandth of a millimeter. Courtesy G.I.M McFadden.

capable of occupying new niches and invading older ones, than their presymbiotic ancestors had been.

So powerful is the advantage of symbiosis that it has happened repeatedly. There are present-day organisms that have undergone repeated symbioses, and some of these have taken place quite recently. Their cells are like Chinese boxes, consisting of layer upon layer of symbiotic organisms.

Consider, for example, cryptomonads—ovoid, single-celled organisms that propel themselves through the water with two rapidly beating flagella. Most live in fresh water, but some are marine. Some cryptomonads are photosynthetic, but others have no sign of chlorophyll.

The more that is learned about these tiny floating creatures, the more complicated their story becomes. They seem to have started out as nonphotosynthetic, animal-like protozoa, but since that time different cryptomonad species have repeatedly lost and gained the ability to photosynthesize. The methods by which they have done so have been dramatic. Some species have swallowed single-celled green algae, which now live inside their cells. Others have engulfed red algae, which were also free-living single-celled photosynthetic organisms in their former lives (Figure 9.5).

EVOLUTION BY COMMITTEE

The story of these greedy cells becomes even more complicated when we remember that the ancestors of the red and green algae that they so eagerly devoured had much earlier acquired their own chloroplasts by engulfing cyanobacteria. A typical photosynthetic cryptomonad has a set of genes on its chromosomes and a second set—descended from bacteria—in its mitochondria. It has a third set, much damaged and inactivated, inside the degenerated nucleus of the red or green alga that it has swallowed. The alga in turn has its own set of mitochondria, contributing a fourth set of genes. And finally, there is a fifth set, contained in the chloroplast inside the alga! This jostling collection of genes works together remarkably well, though there is now, of course, no trace of all those failed evolutionary experiments that must have taken place but that did not lead to a successful cryptomonad.

Ancient Symbioses?

Cryptomonads are by no means the only organisms that have benefited from such complex and continuing symbiotic events. We can catch tantalizing glimpses of much earlier symbioses. Margulis and others have suggested that symbioses gave the evolutionary process during those early days unparalleled flexibility and power, although it must be cautioned that the evidence for such symbioses is fragmentary at best.

Chloroplasts and mitochondria still carry their own genes, though the number is much reduced since the time when their ancestors were free-living. But if there were earlier symbioses that took place during the very early history of eukaryotic cells, they have left no such convenient genetic clues—the genes of those earlier symbionts have apparently all been lost or transferred into the nucleus. We must therefore search for clues elsewhere.

Margulis suspects that the ubiquitous structures known as *microtubules* were once parts of free-living creatures. These tiny protein tubes are responsible for much of a cell's structure and internal movement, including the migration and sorting of its chromosomes. She hypothesizes that these microtubules might have descended from a free-living organism that had the properties of a *spirochete*. Spirochetes are little spiral bacteria found everywhere in our environment. Some of them are responsible for diseases such as syphilis and yaws.

Margulis is encouraged in this supposition by the observation that several single-celled protozoa have actually made slaves out of spirochetes. They provide places on their outer surfaces for the spirochetes

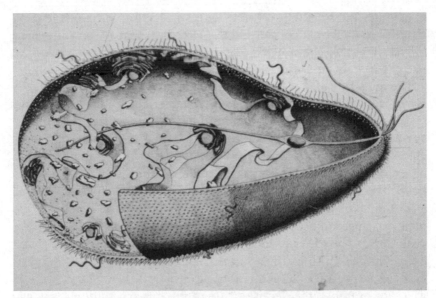

FIGURE 9.6 Picture of *Myxotricha*, the termite gut symbiont, showing its spirochetes, its surface bacteria, and the bacteria living inside it that actually digest the cellulose that the termite eats. (Diagram by Laszlo Meszoly, courtesy Lynn Margulis)

to nestle and even provide a place for other bacteria that feed the spirochetes a trickle of ATP. Tied down in these little niches, like oarsmen chained to benches on a Roman galley, the enslaved spirochetes wave back and forth and propel their hosts smoothly here and there (Figure 9.6).

Unfortunately for Margulis's pretty theory, spirochetes do not seem to make the protein called tubulin, the major constituent of microtubules. If some kind of free-living creatures really did invade our cells in the distant past, providing them with microtubules in the process, it seems that they were not spirochetes.

The possibility of a symbiotic source for microtubules has by no means been ruled out, however. One way to find out would be to look for free-living prokaryotic relatives of the original microtubule-like cells that might have taken part in the symbiosis. Such cells might be living anywhere in the soil, in deep rocks, in the ocean, or in the sludge of an estuary. It should be possible to use the highly sensitive PCR technique to comb samples of soil and mud for genes that are very similar to the tubulin gene. The search would have to be confined to samples unlikely to contain eukaryotic cells, because all such cells contain tubulin.

The spirochete idea and various other speculative possibilities for early symbiosis that have been suggested by Margulis and others have usually been dismissed by the scientific community. History may be repeating itself here—remember that the symbiotic origins of mitochondria and chloroplasts were also dismissed. Because those important structures have turned out to be the result of symbioses, it seems highly likely that traces of many earlier symbiotic events remain to be discovered.

These three evolutionary processes—hopeful monsters, massive genetic exchange, and symbioses—go a long way toward explaining how evolution can be sped up, allowing some organisms to achieve tremendous complexity since life's simple beginnings. But other dramatic events, such as waves of extinction, have shaped the history of life as well. For example, the demise of the dinosaurs 65 million years ago opened up many ecological niches for a wide variety of mammals, including our own ancestors. A hundred and forty million years earlier, an even more massive extinction gave an evolutionary shot in the arm to the ancestors of the dinosaurs. Could such events also provide clues to the appearance of early life?

As we go further back in time, the effects of mass extinctions become hard to interpret. To understand their influence, we must apply what we have already learned about symbioses and massive genetic exchanges. Only then can we make sense out of the earliest history of the Earth.

The Doolittle Event

In the late 1990s, biochemist Russell Doolittle of the University of California at San Diego set the scientific world on its ear, something that he takes great delight in doing. He has found data that, by one interpretation, suggest that somewhere between 2.0 and 2.5 billion years ago, an event occurred that was so dramatic that it could easily have plunged the Earth back into lifelessness. In the process, he has crossed swords with the dean of the early fossil record, Bill Schopf of the University of California at Los Angeles.

According to Schopf's and others' interpretation of Doolittle's results, it is possible that virtually every organism living at that time suddenly died, leaving only a few survivors of a single species to repopulate the planet. This extinction event seems to have happened late in the Earth's history, when the atmosphere was no longer reducing and the puny supply of ingredients from space had diminished to virtually nothing. This means that if that last species had died, then life might never have reappeared. In this frightening scenario, all life on Earth came within a whisker of permanent extinction. We call this brush with total extinction the Doolittle event.

By another interpretation, although the event could indeed have been dramatic, it was more stretched out in time; there was no point at which it brought all of life to the edge of extinction. In this second interpretation, an important role is played by reticulate evolution, the exchange of genes across different branches of the evolutionary tree. An equally important role is played by local rather than massive extinctions.

As we try to decide between these possibilities, we will discover clues to the very early evolution of life. We will begin to understand why the tree of life itself is so confusing and inconsistent.

Doolittle first announced his results in 1996. He examined fifty-seven different families of proteins from a wide variety of animals, plants, and bacteria, along with a few archaea. Within each of the families, the proteins had diverged from a common ancestor that lived billions of years ago, gradually accumulating mutational differences in the process. By examining these differences, Doolittle could trace the historical relationships among them.

At present, a protein in humans is only a little different from the corresponding protein in horses, since the common ancestor of humans and horses lived a rather short time ago—perhaps 70 to 100 million years. But the difference between a protein in humans and the corresponding protein in bacteria is far greater, because the common ancestor of humans and bacteria lived billions of years ago.

Doolittle's idea was to try to trace each of these protein families back to the common ancestor genes from which each family arose. His expectation was that it should be possible to follow each family back nearly all the way to the base of the tree of life, perhaps three billion or more years ago. But he found to his surprise that the molecular trail for each family ends abruptly far more recently than that. The genes from humans and bacteria, for example, turn out not to have diverged as far from each other as would be expected if their common ancestor goes back all the way to the beginning of life.

To put a date on the molecular divergences that he saw, he extrapolated back from known—or approximately known—times of divergence that could be inferred from the fossil record. Some of these dates are fairly firm—many mammalian lineages, for example, can be traced back a mere 65 million years, to the end of the age of dinosaurs and the beginning of the great mammalian radiation. Other divergence times, such as those marking the split between major groups of multicellular animals, are more difficult to pin down.

For example, the time of the split between two great superphlya of complex animals known as the annelid-arthropods and the echinoderm-chordates is difficult to infer from the fossil record. In one direction, this split has led to the segmented worms, the mollusks such as squids and clams, and the huge group of arthropods that includes all the insects. In the other direction it has led to the echinoderms—sea urchins and their relatives—and to the animals with backbones such as ourselves.

The earliest fossil evidence for this split between the superphyla has been found in rocks of the early Cambrian period, about 570 million years ago. From these fossils, however, it is obvious that the split was already well advanced, because the fossils clearly fall into one group or the other. But it is here that the fossil trail runs out, for no representatives of these groups have been found in rocks older than the Cambrian.

Because the earliest history of this split is missing from the fossil record, we know nothing about the creature that must have been the common ancestor of these two highly divergent groups of organisms. Presumably it was small and undistinguished in appearance and crawled around on (or in) the mud of the sea bottom many millions of years before the start of the Cambrian. Traces of such tiny, soft-bodied creatures are relatively rare, even from periods when fossils are plentiful, and very few fossils of any kind have been found in Precambrian rocks. If a fossil of this little animal were to be found today, we would probably not recognize how important it was.

Doolittle found that when he extrapolated his molecular data back using well-established dates from the fossil record, he obtained a time of about 670 million years ago for this superphyla split. This provided a satisfactory cushion of 100 million years between the earliest fossils and the molecular divergence. It also explained why the members of the two superphyla had become so clearly different by 570 million years ago. A hundred million years gave plenty of time for evolutionary divergence to take place before the first creatures that showed evidence of the split began to leave fossils behind.

Encouraged by this, he extrapolated even further back, drawing together all the divergent threads of living organisms, including far flung bacteria and archaea that are scattered across the remotest branches of the tree of life. When he did so, he found that he could go no further back than 2, or perhaps at the most 2.5, billion years. All the protein families, no matter how rapidly or slowly the genes in each family had evolved, tended to converge on their common ancestors by that date. It really looked as if every modern-day organism, ranging from bacteria through plants and animals, had a single common ancestor at around that time.

But this was at least a billion and a half to two billion years after the first appearance of life! The obvious explanation, and the one that Schopf immediately quarreled with, was that there must have been some terrible upheaval, far more severe than the asteroid impact that drove the dinosaurs to extinction 2 billion years later. That event would have wiped out as much as 2 billion years of evolutionary history, killing almost all the immense armies of organisms that inhabited the planet at the time. It left only one solitary branch of survivors, from which all the organisms living on the Earth today are descended. The genes of these few survivors began to diverge again as the planet became repopulated, eventually giving rise to the genes of the present-day organisms that Doolittle used in his survey.

Death and Resurrection

Bill Schopf knew that such a catastrophe was unlikely, for the fossil record showed continuity through the entire period surrounding the presumptive Doolittle event. Further, Schopf had himself collected much evidence that blue-green algae (or cells essentially indistinguishable from them) existed at that time and had probably evolved long before.

The fossil record appeared to show that blue-green algae had already evolved by 3.5 billion years ago. Indeed, they seem to have changed remarkably little right down to the present. Because of this apparent stasis, Schopf has dubbed the blue-green algae the grand champions of slow evolutionary change. Surely, therefore, the genes of blue-green algae should be traceable far back before that hypothetical event.

Doolittle countered with his molecular evidence that the genes of blue-green algae have undergone no more divergence than those of many other groups of bacteria. The divergence could therefore be dated back only to the Doolittle event, although the blue-green algae might be older than that and might have given rise to other kinds of organisms after the event.

The geological record is equivocal about the Doolittle event. As we saw earlier, geological evidence indicates a massive glaciation peaking at 2.2 billion years ago, but this seems to have had little effect on life at the time. There is also evidence for a meteorite impact, the vast Sudbury crater in eastern Canada, near Lake Huron, which is nearly one hundred kilometers in diameter, that dates near the time of the event. The bolide that formed this crater, however, crashed into our planet 1.85 billion years ago, somewhat later than the time of the apparent disaster suggested by the evidence from the genes. And the Sudbury event, although huge, was unlikely to have been so devastating that it killed practically every organism on the planet. The bolide was probably about the same size as the one that hit the Earth 65 million years ago and that led to the demise of the dinosaurs. That impact, severe as it was, did not come even close to wiping out all life.

Indeed, rather than some great upheaval taking place between 2.5 and 2 billion years ago, it seems that the Earth was actually settling down after the numerous bolide impacts of earlier epochs. During this whole span of time, the fossil record shows that blue-green algae were even more plentiful and varied than they had been at earlier times. Further, the species living during that period were not very different from those found today.

What really happened? One immediate response to Doolittle's conclusion was that he was wrong to extrapolate back in a linear fashion from recent fossil evidence to the very distant past. Suppose that evolution has gradually sped up as we approach the present. Then, the further back you go, the more slowly genes would have evolved. It would not take much of a change in evolutionary rate to extend his time of ultimate gene coalescence to 3.5 billion years ago instead of 2.5 billion.

Doolittle's counterargument was that all these different gene families were very unlikely to have undergone the same amount of slowing. Moreover, when he picked the slowest- and fastest-evolving members of his gene families, he still got the same date for the earliest gene divergence. But there is another, more powerful argument in his favor that he did not mention. It turns out that the genes actually do provide us with a glimpse of that first billion or more years, the period before the hypothetical disaster of the Doolittle event.

Naoyuki Iwabe of Kyushu University and a group of colleagues have achieved a dramatic insight into the early history of life. They began by looking for pairs of genes that arose by gene duplication so long ago that both members of the pair are possessed by every known living organism. They settled on two pairs that are utterly fundamental to the way that living cells work. One pair specifies different parts of the ATP-manufacturing protein (see Chapter 7). This protein makes ATP in all creatures, from bacteria to humans. The other pair specifies important factors that aid in the manufacture of proteins in the cell, and this pair too is found in every known organism.

Because the genes of each pair are related to each other, they must have had a common ancestral gene. This gene must have become duplicated at some point in the very distant past, so that one gene has become two. Such a duplication is not surprising, for gene duplications are very common in evolution—we carry many families of genes that are related to each other and that must have multiplied on our chromosomes by a process of duplication. Once the duplication occurred, then the duplicates themselves were free to diverge and to become progressively more and more different from each other.

Remember that every organism living today has both members of each gene pair. This means that all the organisms that had only one gene must have gone extinct, leaving only those that carried the duplicated pair. The duplications must have occurred before the Doolittle event that led to the establishment of all present-day organisms. Indeed, Wen-Hsiung Li of the University of Texas has used the amount of divergence between the duplicates to date some of Iwabe's duplication events to more than three billion years ago.

Figure 9.7 shows, in simple form, the history of one of these two pairs of duplicated genes. The dashed line emphasizes the fact that we know nothing about what happened to the original gene from when it first appeared to when it duplicated—how much it evolved and how long it took to do it. All that information has been lost in the mists of the past, perhaps never to be recovered. But we can follow the history of these pairs of genes after the duplication took place, by looking at how much the genes have diverged since their duplication.

Iwabe's results show clearly that the divergence has been very great, so great that the history of these genes extends back before the Doolittle event. We share that history with all the organisms living in the world today. There seems to be no other conclusion than that Doolittle is right. Long after the origin of life, some kind of dramatic

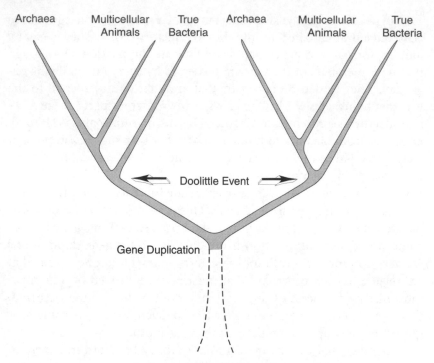

FIGURE 9.7 Iwabe's evidence that much evolution took place before the earliest common ancestor that we can trace back using present-day genes. Every organism now living has two copies on different parts of its chromosomes of certain fundamentally important genes. The duplication event that led to these two copies, as well as some of their subsequent divergence, must have happened before that earliest common ancestor.

event or events took place, involving substantial extinctions and giving rise to all the species that we see today.

If the Doolittle event were some huge catastrophe, even though it somehow left no obvious trace in the geological or fossil record, it must have been truly frightening. If it had been just a little more severe, all living things would have vanished. Life itself hung by a thread. Had the thread snapped, then the Earth might have spent the rest of its long career spinning through space as an utterly dead planet.

Perhaps things were not that bad, however. There is another and less apocalyptic explanation for what might have happened. One clue to this is provided by the archaea, those mysterious organisms often found in extreme environments and that we discussed in Chapter 8.

Catastrophe or Replacements?

The complete sets of genes possessed by five different archaea have recently been sequenced, and more will soon be completed. Some of these newly sequenced genes are members of the gene families that Doolittle had originally used. He found that when he put these new genes into his data set, some agreed with his earlier findings. They had begun to diverge roughly 2 billion years ago. Some, however, had begun to diverge much earlier, perhaps 3.5 billion years ago.

The disparate data are disorienting. How can some archaeal genes be young, and others old? One possible explanation is that archaea are a kind of genetic pastiche, the result of symbiotic events or massive gene exchanges that took place in the distant past.

Indeed, there is growing evidence that this is so. Scientists have suspected for some time that there is something very strange about archaeal genes. Some resemble the genes of creatures like ourselves; others resemble the genes of true bacteria. A third class seems to be extremely different and to have no identifiable relatives.

To make the picture even more confusing, some archaea have borrowed collections of genes from bacteria, and others have not. *Archaeoglobus fulgidus* is an archaeon that lives deep in the ocean, surviving on seeps from underwater oil deposits. It has borrowed fat-processing genes from bacteria, but these genes are not found in any of the other four archaea that have had all their genes sequenced.

So it seems that massive gene swaps have taken place between some archaea and some bacteria. Given this fact, it becomes possible to explain the apparent Doolittle event—and the powerful evidence for an earlier vanished world seen in Iwabe's duplicated genes—without invoking a worldwide catastrophe.

We suspect that the most likely scenario is this. Even though the sparse fossil record shows little evidence of it, the geological record provides strong indications that the Earth was changing dramatically during the period from 2.5 to 2.0 billion years ago. As a result of photosynthesis, free oxygen was beginning to build up in the atmosphere for the first time. Anaerobic bacteria and archaea, most of which even today are poisoned by oxygen, were being driven out of the planetwide ecological niches that they had occupied for so long. The mitochondrial and chloroplast symbioses were taking place, and air-breathing and photosynthesizing organisms were proliferating. There was, in short, ample opportunity for dramatic and sweeping ecological replacements of one group of organisms by another.

Suppose that a large number of such replacements occurred over this span of half a billion years. These replacements would have been driven by symbiotic events and by massive gene swaps, producing new and more powerfully competitive hybrid organisms.

None of these replacements would have needed to have been complete by itself, but the cumulative effect would have been a gradual but complete replacement of the old organisms with new and more vigorously competitive ones. Figure 9.8 shows, in diagrammatic fashion, how this might have happened. The figure shows when the Iwabe duplications took place and the role of gene swaps and symbioses during the time since.

Consider the mitochondrial symbiosis, about which we know the most. There is growing evidence that some genes from the bacterial invader rapidly hopped over into the chromosomes of their new host. Mitochondria were already a very good thing to have, but this new transfusion of bacterial genes may have had equally powerful consequences. The eukaryotes that had mitochondria and all these additional new genes eventually drove those that did not have them to extinction, a process that could easily have taken hundreds of millions of years but that was eventually complete.

Similar and earlier replacements, the result of different selective pressures, would have taken place in the worlds of bacteria and archaea. In each case, the complacent, old-fashioned organisms were replaced by vigorous new hybrids. We may never know the details of these replacements, but they are less discomfiting to imagine than a sudden worldwide catastrophe. Because there is at least a half-billion-year uncertainty in the date of the Doolittle event, such replacements could have taken hundreds of millions of years and still give the impression from our distant vantage point that all present-day life can be traced back to one traumatic event.

Could such an astonishing scenario have taken place? Half a billion years is roughly the same period of time that complex multicellular organisms have existed on the planet. Such a vast span of years would have given plenty of opportunity for the new superorganisms to diversify into many different ecological niches. Since all this happened before the appearance of multicellular life, and both the new and the old

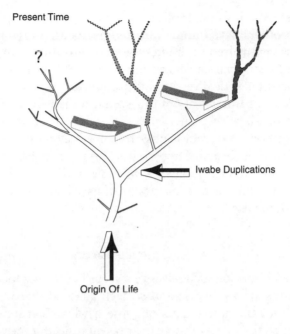

Present Time

?

Iwabe Duplications

Origin Of Life

———— "Old" Organisms

} "NEW" More Competitive Organisms

FIGURE 9.8   A simplified tree showing how some "old" organisms underwent the Iwabe gene duplications, and others later donated genes to some of these lineages. The new, more vigorously competitive species that resulted began to spread, displacing the old organisms that did not carry the duplications. Further donations of genes produced a variety of lineages of these highly competitive species. Eventually, all the old preduplication strains everywhere in the world were replaced by new ones, but some of the genetic diversity of the old strains was saved by the gene transfers.

organisms were very simple single-celled creatures, there would be no obvious trace of these replacements in the fossil record and no near brush with the total termination of all life.

This would explain the pattern that Iwabe saw. All the survivors of these selective sweeps carried Iwabe's duplicated genes, and all the organisms that carried only single copies of the ancestral genes eventually perished. And it would explain the very old archaeal genes that Doolittle and others have found. Various diverged representatives of these genes were involved in different gene swaps during this period. Even though the original carriers of those genes went extinct, at least some of that ancient genetic diversity was preserved.

We feel less worried about this scenario than we do about some terrifying brush with planetwide extinction. It seems likely that the Doolittle event was really a long-drawn-out Doolittle replacement. And this view of life has the happy result, from our perspective, that it emphasizes the immense role of symbioses and genetic exchanges in the evolution of life. Such exchanges must have had a direct bearing on the origin of life itself.

Genetic Anarchy

We cannot imagine such selective sweeps happening today. Present-day organisms are so diverse and so well adapted that it would be very difficult for the descendants of some newly arisen superorganisms to displace all the species alive at the present time. But the world of 2.5 billion years ago was a less complicated place. Sufficiently great advantages, such as a better way to extract energy and thus to increase the speed of cell division, might have been enough for superorganism lineages to spread through many ecological niches. Because anaerobic bacteria were replaced along with aerobic cells, the advantage was probably not just the ability to survive in the presence of free oxygen.

The lineage that ends in a question mark in Figure 9.8 is meant to indicate that we do not have the final word on whether any of the old, pre-Iwabe organisms have escaped down to the present. If they did, then we have yet to find them. Perhaps they have persisted in unusual ecological niches, where they have not been required to compete against those parvenu superorganisms with their advanced properties.

Such survivors might be found among the organisms living in fissures deep in the rock that we described in Chapter 8. As we discussed, the majority of those elusive creatures have been detected only from small fragments of their DNA. They have proved impossible to culture in the laboratory because they multiply too slowly or not at all under any conditions that we have provided them. But when these difficulties are overcome, it seems very likely that new glimpses of that long-vanished world from before 2.5 billion years ago will soon be found.

Until these evolutionary relics are found, we can only guess about the genetic exchanges and symbioses that might have happened before the Doolittle replacement. As we mentioned at the start of this chapter, Carl Woese, the discoverer of the archaea, has recently proposed that there was so much lateral gene transfer soon after life's origins that any pattern in the earliest parts of the tree of life has been completely lost.

As a result, Woese says, it is not possible to trace individual lineages of organisms back to the last common ancestor. The earliest cells, he suggests, were actually aggregated communities of very simple protoorganisms, each having some of the properties needed to allow the entire community to survive. Each part of such a community would have contributed an important talent, but the whole would have been greater than the sum of its parts. Symbioses and massive gene exchanges could have played an important role in all this, by combining groups of genes that conferred very different properties. And these characteristics could have been mixed and matched repeatedly, giving rise to cooperative groups of protoorganisms that had very different properties.

We agree with Woese that there could have been a long period, probably extending back to the appearance of the first primitive protobionts, during which gene-swapping, parasitisms that evolved into symbiosis, and all sorts of other things went on. We can think of these early living entities as having gone through a period of experimentation rather like the sexual revolution of the sixties! But it remains to be seen whether all of the structure that was possessed by the early evolutionary tree was lost in the process. More likely, as we sequence more genes and explore some of the more remote branches of the tree of life, we will discover evidence for those early gene swaps and symbiotic events.

The prospect of this has molecular evolutionists very excited. These new studies are not only casting light on very early stages in the evolution of life, but are also helping to explain one of the enduring mysteries of the evolutionary process itself—how some living organisms have become so complex and how this complexity might have arisen rapidly through symbiotic events.

It is possible to return to the scenarios we painted in Chapter 4, in which we tried to reconstruct the mechanisms by which complex sets of molecules could have arisen even before the appearance of life. As we suggested, several different kinds of genetic material might have been present, each coding for some primitive but nonetheless necessary factor for early metabolism, or for some information necessary for gene duplication. Each bit of information, unable to exist by itself, would have depended on other bits of information, perhaps coded by molecules that were chemically somewhat different. Symbiotic events would have brought those protobionts together, giving them a great competitive advantage.

The ultimate test of the top-down approach to the origin of life is to see how many traces of these early events can still be found in our genes, and whether we can reconstruct some of them in the laboratory. This was hard to do when only a scattering of genes were known from any organism. But now the pace of gene sequencing is increasing exponentially, with hundreds of millions of new bases of DNA being sequenced every year. More and more species (including ourselves) will soon have their genes completely sequenced. The shape of the earliest events in the evolution of life, which we can now only glimpse in outline, will grow ever plainer in the decades to come.

LIFE ELSEWHERE

We, certainly, cannot say how life originated and developed on other celestial bodies under the conditions peculiar to them. But it stands to reason that the organisms forming in the process of biological evolution must differ essentially from the terrestrial animals and plants since it is the environment that forms life.

Alexander Oparin, Life in the Universe, 1961

IN 1655, THE DUTCH ASTRONOMER CHRISTIAAN HUYGENS (Figure 10.1) discovered Titan, the giant moon of Saturn. He did it by peering through a new, state-of-the-art telescope that he had made himself. He was awestruck by the beautiful, distant mini–solar systems of moons that he saw revolving around Jupiter and Saturn, and wondered if "the wise Creator has disposed of all his Animals and Plants here [on Earth], has furnished and adorn'd this Spot only, and has left all those worlds bare and destitute of Inhabitants, who might adore and worship Him?"

Huygens's question was a heretical one at the time, and of course unanswerable. But now, as we have begun to explore our Solar System rather than just observe it, the question is being raised again. The satellites of the massive gas giant planets Jupiter, Saturn, and Uranus were formed at the same time as the planets themselves were condensing from the great disk of gas and dust that surrounded the Sun. There are so many of these moons, and they are all so different from one another, that it will take many human generations to explore them all. And life may exist on more than one of them.

Planet of Smog

Titan is bigger than Mercury and is the second largest moon in the Solar System (the biggest is Jupiter's Ganymede). It is shrouded by a dense, reddish haze of atmosphere, which makes it unique among the Solar System's moons. We know that there is a great deal of frozen wa-

FIGURE 10.1 Christiaan Huygens (1629–1695). From P. Lenard, *Great Men of Science: A History of Scientific Progress* (London: G. Bells and Sons, Ltd., 1954).

ter on Titan, because the entire moon has a density only slightly greater than water ice. But we cannot glimpse the ice, or indeed any part of Titan's surface, through the smoggy clouds.

Titan's atmospheric pressure is 60 percent greater than that of the Earth. It is primarily made up of nitrogen, like the Earth's atmosphere, but there the resemblance ends. Astronomer Gerard Kuiper first studied the atmosphere in detail in 1944; he detected the presence of methane. Other simple hydrocarbons and hydrogen cyanide were detected later, and this mix of energy-rich reducing gases was confirmed when *Voyager 1* flew by in 1980.

Remarkably, Titan's atmosphere resembles the reducing atmosphere proposed by Harold Urey for the early Earth. To emphasize the point, Carl Sagan and his co-workers experimented with a simulated Titan atmospheric mixture, which they exposed to ultraviolet radiation. They found that amino acids and a variety of other simple organic compounds were produced, along with a reddish polymer similar in color to the haze that cloaks the moon itself.

Titan is probably a giant prebiotic organic chemical factory. But it is a chilly one: Surface temperatures are −178°C, only 95° above absolute zero. One reason for this is that the thick clouds act as an anti-greenhouse layer, reflecting back most of what little heat manages to reach

THE SPARK OF LIFE

Titan from the distant Sun. Nevertheless, some researchers have suggested that there may be oceans or lakes on Titan—not liquid water oceans, but oceans of liquid ethane or other simple hydrocarbons. Another possibility is that the surface is made up of a kind of slushy mixture of water and organic compounds, something that Tobias Owen of the University of Hawaii has dubbed "the primordial ice cream."

This slush would be an ideal site for the synthesis of complex organic compounds such as adenine. You will recall from Chapter 2 that Miller and his colleagues have made adenine from frozen spark-discharge solutions. Might there be quantities of adenine on Titan? What else might there be?

We may find out in 2004. After a 6-1/2-year tour through the Solar System, the *Cassini* spacecraft will reach Saturn's orbit and begin a detailed study of the planet, its rings, and its moons. Then in November of that year, the *Cassini* orbiter will release the Huygens probe, which will coast toward Titan for three weeks. Slowed by the outer reaches of the atmosphere, the probe will make its final descent by parachute.

After the parachute deploys, forty kilometers above the surface, an ingenious device will make the probe spin slowly, enabling it to take panoramic pictures of the descent through the clouds as well as pictures looking straight down. Instruments carried on board will provide information about the components of Titan's atmosphere and the amount of sunlight that manages to reach the surface through the dense cloud cover. Aerosol particles in the atmosphere will be heated in a tiny oven, and the resulting gas will be analyzed, allowing detection of organic compounds. After the probe lands on the mysterious surface (Plate 10), it is scheduled to broadcast data for up to an hour.

The Huygens probe will give us a glimpse of a far-off world on which conditions are very much like those thought to have existed on the early Earth. There are obvious differences, of course—Titan is very cold, but the early Earth must have been much warmer. Still, it is possible that tidal heating has produced localized volcanic eruptions on Titan's surface, and those warmed-up patches could be cradles for life. We will probably not find a better match to the young Earth until we venture out among the stars.

Listening for Life

Ever since its triumphant expeditions to the Earth's moon, NASA has often seemed to be an agency groping for a mission. This is soon likely to change, and to change dramatically. Riding on a wave of astounding new scientific discoveries about the Solar System and our immediate galactic neighborhood, NASA's chief administrator, Daniel Goldin, has become an enthusiastic supporter of the search for extraterrestrial life. He has announced that one of the big goals for the agency in the coming decades is to look for life elsewhere in the Solar System and beyond.

There is no doubt, as Goldin fully realizes, that the discovery of extraterrestrial life will change utterly our perception of our place in the universe. Of course, we may find that there is no life anywhere else in the Solar System itself, and even if we do find life, it probably will merely consist of very simple bacterial cells. But suppose that in the next decade or two, signs of life are detected on a planet or moon orbiting a nearby star, in the form of a strong signal indicating substantial amounts of free oxygen in its atmosphere. Such a world would certainly be an abode of life, and the life it carries would have a high probability of being complex and multicellular.

The goal of reaching that world and exploring it, and of finding new and ever-swifter ways to do so, would galvanize the entire human race. NASA would no longer need to plead endlessly for little driblets of funding. It would be the leader in the most exciting adventure ever undertaken by our species.

An encouraging indication of the public's growing interest is that Goldin's espousal of this daring new direction for NASA was not greeted with a barrage of ridicule. Only two decades earlier, in 1978, NASA had proposed to give $14 million to support SETI, or Search for Extraterrestrial Intelligence. The idea, a project to search for radio signals from extraterrestrial civilizations, had been developed by Frank Drake, a radio astronomer at the Green Bank observatory in West Virginia. Senator William Proxmire of Wisconsin immediately gave the idea his dreaded Golden Fleece award, which was quite enough to kill any possibility of funding.

Drake, stung by Proxmire's award, struck back by nominating him for membership in the Flat Earth Society. But it was too late. For the next three years, Proxmire used his position on the Senate Appropriations Committee to ensure that there was no further funding for such harebrained schemes. The senator was persuaded to withdraw his opposition in 1981, but only after being lobbied strenuously by Carl Sagan.

So terrified is Congress of appearing ridiculous that even now, NASA gives out no funds for such searches. Much of the funding currently comes from the privately sponsored SETI Institute, which was founded by Drake.

The search is a vast one. Drake had embarked on a tentative attempt in 1960, a scheme he called Project Ozma. It could only scan a few radio frequencies and was confined to a handful of nearby stars. Not surprisingly, he found nothing.

The ability to search the skies for extraterrestrial signals has advanced dizzyingly in the subsequent forty years, powered by rapidly improving radio technology and computers. One of the most interesting current efforts is being run by Dan Werthimer, an astronomer at the University of California, Berkeley. He has hitched a sensitive detector to the giant radio telescope that occupies a bowl-shaped valley at Arecibo in Puerto Rico. The detector obtains a free ride as the Earth rotates and the telescope ponderously scans the heavens. As it rides along, it can scan 100 million different frequencies, all near the hydrogen absorption line, at which there is little other natural radiation.

The detector gathers so much data that Werthimer has devised an ingenious scheme for analyzing it. Internet surfers can download (from SETI@home.com) a screen saver that examines a small piece of the data while their computers are not being used. So far, some 700,000 people in 220 different countries and territories have participated in the program, contributing 18,000 years of computer time. Alas, although some unusual signals have been turned up, there is nothing so striking that it suggests the activities of an alien civilization.

The probabilities underlying the search are daunting in their unlikelihood. To calculate the possibility of Ozma's success, Drake used the following simple equation to estimate the number of detectable, communicating civilizations (N_c) in our galaxy, the Milky Way:

$$N_c = R \times f_p \times n_e \times f_l \times f_i \times f_c \times L$$

R is the number of stars formed in our galaxy each year, f_p is the fraction of these stars that have planets, n_e is the number of planets in each of these solar systems that can support life, f_l is the fraction of these planets that in turn actually develop life, f_i is the subset of this fraction of planets that develop intelligent life, f_e is the proportion of these where intelligent species develop some observable technology such as radio, and finally L is the period of time, in years, during which such civilizations could be detected before they go extinct or develop communication capabilities beyond our detection technologies.

There are huge uncertainties in all the parameters in this equation, except for perhaps the first, which astronomers can estimate from the observed rate of star formation in nebulas and star clusters. It is this rate of star formation that governs the number of possible civilizations in the galaxy at any one time, not (as is often asserted) the total number of stars in the galaxy! The best estimate at the moment is that R is about 10 stars per year.

Optimists would suggest that all the other terms in the equation, except for the last one, are roughly equal to 1. That is, every star has a planet or moon on which life can develop, and intelligent life is likely to follow sooner or later. If we suppose, as Drake did, that the last term is approximately 10,000—that a civilization we could detect persists for 10,000 years—then at any given time, there should be 10 x 1 x 1 x 1 x 1 x 1 x 10,000 = 100,000 civilizations in the galaxy. The means that one star is a million (there are about 100 billion stars in the Milky Way) might carry such a civilization in our galaxy alone! And, there are a lot of other galaxies in the vastness of the universe.

The average distance between such civilizations in the Milky Way would be less than a thousand light-years, but most of these civilizations would presumably be found in the thickly star-strewn galactic nucleus or in the dense star clusters that orbit the galaxy's outer reaches. Among the thinly scattered stars of the galactic arm where our own Sun is located, the distances between civilizations are likely to be very much greater. Even so, such distances could conceivably be bridged by the various giant radio telescopes that are in the planning stages.

The last few terms of Drake's equation are the most uncertain, because they deal with the development of intelligent life and how long it might persist. What excited Goldin, and is exciting astronomers and planetary scientists around the world, is the growing possibility of short-circuiting the equation. It is now becoming possible to search for planets that simply have life, even if they have not developed to the point of possessing intelligence.

This changes the Drake equation dramatically. The highly problematical terms f_i and f_c, which measure the likelihood of intelligent life and the appearance of a detectable civilization, can be jettisoned. And the last term, L, becomes the number of years during which life itself, not just intelligent life, can persist on a planet. On Earth, L is over 3.5 billion years and growing.

In its short-circuited form, the Drake equation yields the number of places (N_{life}) where life exists in the Milky Way. And N_{life} is very large, certainly in the millions even under the most pessimistic assumptions, and reaching into the billions if we assume that the other factors after R are close to unity. The distance between such planets becomes on the order of a light-year or even much less. Even in our sparse region of the galaxy, there may be only a few light-years separating planets that have life. With odds like these, Goldin is making a very safe bet indeed!

The Properties of Extraterrestrial Life

What would the life on other worlds be like? Must it be made up only of carbon-based compounds? H. G. Wells, among others, thought otherwise. Writing in the *Pall Mall Gazette* in 1894, three years before he published *The War of the Worlds*, Wells imagined that "the elements silicon and aluminum might play the role now played by carbon and nitrogen in the chemical processes that underlie the phenomena of life." He went on to scold those who thought only of carbon-based life: "It is narrow materialism that would restrict sentient existence to one series of chemical compounds, . . . and the conception of living creatures with bodies made up of the heavier metallic elements and living in an atmosphere of gaseous sulfur is by no means so incredible as it may, at first sight, appear."

Notwithstanding Wells's broadside, chemists were learning that there are good reasons why life on Earth is based on carbon compounds. Carbon is one of the most abundant elements in the universe. It can readily form chemical bonds with many other elements, especially nitrogen, oxygen, hydrogen, sulfur, and phosphorus. As a result, a vast array of carbon-based compounds is possible, and a huge number of these have been synthesized in the laboratory—as evidenced by the multivolume compendium *Beilsteins Handbuch der Organische Chemie* (see Chapter 2). Life itself has been far more inventive than organic chemists, for a vast number of organic compounds found in living cells have so far defeated the capabilities of even the most talented synthetic chemists.

Although silicon-based life has been suggested as a possibility, silicon can bond with only a few other elements and does not form a wide variety of diverse molecules. Silicon-based life would have to be very different from carbon-based life, perhaps drawing on silicon's remarkable electrical properties. But it is not obvious how silicon-based information-carrying molecules, and the silicon-based enzymes necessary to carry out their replication, could arise. Of course, the silicon chips in our computers store vast amounts of information, but these are human constructs, not natural ones, and the chips are incapable as yet of making copies of themselves.

Nonetheless, we should probably keep an open mind. As we explore the vastness of the universe, perhaps we will find that the carbon-only life view is too narrow. H. G. Wells was probably the most accurate prophet of the future who ever lived—he not only predicted atomic bombs in 1913, but predicted how they would work. We may yet encounter Wells's "metallic" organisms!

Is Life Unique to Earth?

Speculation about extraterrestrial life is probably as old as human thought itself. Aristotle assumed that the Earth was unique and different from the heavens in its very nature, but the atomists of the Epicurean school thought that the various bodies in the universe were all made up of similar bodies—and therefore presumably capable of supporting life. Aristotle's view prevailed for a long time, supported by a medieval Church that was anxious to uphold the uniqueness of creation. But with the emergence of modern science at the end of the Middle Ages, Aristotle's notion of singularity began to be questioned.

Giordano Bruno was one of the first to dare to inquire whether we are alone in the universe when he wrote, in *De l'infinito, universe et mondi,* "There are innumerable Suns and innumerable Earths, which revolve around their Suns, as our seven planets revolve around our Sun. These worlds are inhabited by living creatures." For this and other heresies, as we saw in the introduction, Bruno suffered a terrible fate.

Bruno's ghastly end did not deter his contemporaries from asking similar questions. In 1621, the clergyman Robert Burton, in *The Anatomy of Melancholy,* asked, "Why may we not suppose a plurality of worlds?" and wondered whether many might not be habitable. Burton's theme was echoed later in the same century by the Frenchman Bernard le Bovier in his widely read *Conversation on the Plurality of Worlds:* "For what can more concern us, than to know how this world which we inhabit is made; and whether there be any other worlds like it, which are also inhabited as this is?"

As Steven Dick has elegantly summarized in his 1996 book *The Biological Universe*, the concept of many worlds besides our own was the topic of highly emotional and polarized debates in the first part of the twentieth century, often pitting biologists against astronomers. One of the giants in biology was Alfred Russel Wallace, who was the codiscoverer with Darwin of the principle of natural selection. He argued in his 1903 book, *Man's Place in the Universe,* that, considering the unlikelihood of the conditions that have led to the appearance of life on Earth, "it seems in the highest degree improbable that they can all be found again combined either in the Solar System or even in the stellar universe."

Sir Harold Spencer Jones, Astronomer Royal at the Royal Greenwich Observatory in Sussex, England, was much more optimistic. In his 1940 book, *Life on Other Worlds,* he reached the opposite conclusion from Wallace, asserting that it was "inherently improbable that our small Earth can be the only home for life."

Given the continued fascination with this topic, it is not surprising that as the planets and moons of our Solar System were observed with ever-more-powerful telescopes, there were repeated reports of possible signs of extraterrestrial life. Most, but not all, of these claims were made about our closest neighbors in the Solar System: our own Moon and the planets Mars and Venus.

Moon Men

When Galileo first observed the moon with his newly invented telescope in 1610, he was startled to see valleys, mountains, and flat regions that appeared to be seas. Johannes Kepler quickly seized upon this apparent resemblance to the Earth and imagined not only that life existed on the moon but that some of the lunar features were the work of intelligent inhabitants.

The British astronomer Sir William Herschel, in a 1780 letter to Astronomer Royal Nevil Maskelyne, wrote, "Who can say that it is not extremely probable, nay beyond doubt, that there must be inhabitants on the Moon of some kind or another?" Four decades later, the German astronomer Franz von Paula Gruithuisen claimed to have discovered a lunar city made up of "a collection of dark gigantic ramparts."

Public interest in the possibility of lunar life climaxed in the late summer of 1835 with a series of popular articles by Richard Locke that were published in the *New York Sun*. Locke reported that the British astronomer Sir John Herschel (son of William Herschel), while observing the moon at the Cape of Good Hope in South Africa, had discovered bizarre, winged, batlike creatures swarming over it (Figure 10.2). Locke wrote, "Certainly they were human beings. They averaged four feet in

FIGURE 10.2 A portion of the lithograph "Lunar animals and other objects" published as part of a story in the *New York Sun* in August 1835 on lunar life as supposedly observed by John Herschel. (Courtesy the Library of Congress)

height, were covered, except on the face, with short-glossy, copper-coloured hair, and had wings composed of a thin membrane."

The circulation of the *Sun* soared as New Yorkers devoured the news of these exotic lunar beings. A women's club in Massachusetts was so impressed that they wrote to Herschel asking how to contact the lunar bat-men so that they could convert them to Christianity. The story rapidly spread to Europe, where it became the subject of a heated debate sponsored by the Academy of Sciences in Paris.

Herschel, a hemisphere away at the time, was unaware of this idiotic story when it first appeared. As soon as news of it reached him, he denied it furiously, but it was too late. This cruel and pointless hoax had a devastating effect on his career. In 1839 he wrote to a friend, "I have made my mind to consider my astronomical career as terminated."

Speculations about life on the moon did not end with the Herschel affair, but continued well into the twentieth century. In the 1920s and 1930s, the American astronomer William Pickering claimed to have observed a series of strange dark moving spots in the lunar crater Eratosthenes. The spots moved with an "average speed of 6 feet a minute," and he attributed them to "small animals, most likely swarms of insects." We can speculate unkindly in turn that they were more probably small floaters in Pickering's eyes.

Even during the period directly preceding the Apollo missions, microbial lunar life was still considered a possibility. As *Apollo 11* returned with the first lunar samples, elaborate quarantine precautions were put in place to try to ensure that no lunar organisms would escape and infect the Earth.

The precautions were not particularly effective, alas. As the capsule bobbed in the ocean after landing, the hatch was opened just a crack by frogmen, who stuffed biological isolation garments into the opening. The astronauts donned these garments before emerging. Unaccountably, however, the planners overlooked the puffs of moon dust from the capsule's grit-covered interior that emerged along with the astronauts and dispersed unnoticed into the ocean. It apparently occurred to nobody involved that the odds that lunar organisms might attack the ocean's myriad of different species were far higher than the odds that they might attack us. Luckily for our planet, the existence of lunar life was finally put to rest after careful analyses of the Apollo samples showed that the moon was indeed a sterile, inhospitable place.

Venusian Life

Ishtar, Aphrodite, Phosphorus, Venus. These are just a few of the names that ancient civilizations gave to the beautiful heavenly body that periodically appears in either the evening or the morning sky. Because Venus is so similar in size to Earth, it was thought to be teeming with earthlike creatures.

In the late 1700s, when William Herschel trained one of his magnificent new twenty-foot-long telescopes on Venus, he saw a planet enshrouded in clouds. Herschel believed that the cloud cover kept the planet's surface cool, sheltering it from the fierce sun. Clouds on Earth were made of water vapor, so he assumed that water was also abundant on Venus. All this suggested to Herschel the possible existence of life, even intelligent life.

This view was reinforced as more extensive observations of Venus were carried out in the nineteenth century. In 1870, R. A. Proctor wrote that "Venus bears a more striking resemblance to the Earth than any other orb within the Solar System. . . . [T]he evidence we have points very strongly to Venus as the abode of living creatures not unlike the inhabitants on Earth." In 1918 Svante Arrhenius imagined Venus as "dripping wet . . . no doubt covered by swamps," and that low forms of life existed there, "no doubt belonging to the vegetable kingdom."

In 1932, abundant carbon dioxide was detected in the Venusian atmosphere. It was soon realized that the thick clouds on Venus were not made of water vapor and that the planet was in fact bone dry, not the swampy place imagined by Arrhenius. The dense carbon dioxide clouds were soon shown to generate intense greenhouse heating of the Venusian surface, making it so scorching hot that life as we know it could not exist.

In the 1951 revision of his book *Life on Other Worlds*, Sir Harold Spencer Jones summarized the possibility of life on Venus: "The whole planet is a desert. Intense gales blow perpetually over the surface and the yellow dust is carried high into her atmosphere. The surface is consequently steeped in a Stygian gloom. The heat is intense. There is no vegetation of any sort. Venus, then, is a world where life is totally out of the question."

This picture of a hellish Venus did not restrain Carl Sagan and Harold Morowitz from proposing in 1967 that life might still manage to exist in Venus's atmosphere. They argued that life may have appeared at a time early in the history of the planet, when environmental conditions were more favorable than those that exist today. As carbon dioxide built up in the atmosphere and the surface became unbearably hot, life abandoned the surface for the cooler, upper cloud layer.

Sagan and Morowitz imagined a kind of living hot-air balloon, full of hydrogen produced from the photosynthetic splitting of water. The organisms would thus remain suspended at a fixed altitude in Venus's atmosphere, where the temperature was just right for their survival. However, after the discovery in 1973 that the clouds of Venus were filled with corrosive sulfuric acid droplets, any possibility of Sagan's bizarre Venusian life was eliminated.

Martians

The possibility of life on Mars is an entirely different story. In fact, as we start this new millennium, the debate about life on Mars continues. In the coming decades, the search for evidence of life on Mars will be a central focus of the space programs in both the United States and Europe. Spacecraft are scheduled to probe the Martian surface and return samples to Earth for detailed, state-of-the-art analyses.

Speculations about intelligent life on Mars exploded after the Milanese astronomer Giovanni Schiaparelli reported in 1877 that he had seen *canali* on Mars. The word can mean either canals or channels in Italian—Schiaparelli seems to have meant the latter, but his translators assumed he meant the former. These "canals" were soon suspected of being the products of an advanced form of life.

The possibility of intelligent Martians was embraced by the public and some of the scientific community as well. The canals helped to inspire H. G. Wells's *War of the Worlds*. At the same time that Wells's book appeared in 1897, there was a flurry of reports on unidentified flying objects (UFOs), which some journalists suggested might be Martians trying to visit Earth.

A Mars aswarm with intelligent beings was championed by Percival Lowell in his 1906 and 1908 books *Mars and Its Canals* and *Mars as the Abode of Life*. According to Lowell, his observations showed that the Martian canals were far too regular in appearance to have been created by natural processes.

Lowell had his share of critics, however. One was Alfred Russel Wallace, whose 1907 book *Is Mars Habitable?* meticulously set out what was known about the surface of Mars at the time. He concluded, "The first essential of organic life—water—is non-existent. . . . Mars, therefore, is not only uninhabited by intelligent beings such as Mr. Lowell postulates, but is absolutely UNINHABITABLE."

As the twentieth century advanced, the presence of Lowell's canals was eventually disproved, but the possibility of primitive Martian life, especially some sort of vegetative life, continued to have widespread support in parts of the scientific community. At the dawn of the space age in the late 1950s, William Sinton used the powerful tool of infrared spectroscopy to match some of the absorption features he had observed on Mars with the spectrum of absorption that is characteristic of terrestrial plants. As the *Mariner IV* spacecraft was launched toward Mars in 1964, many scientists felt that humankind was finally on the verge of discovering extraterrestrial life.

This optimism was soon shattered in July 1965 as *Mariner IV* swung by Mars and took the first close-up photographs of another planet. The pictures showed a dry, apparently dead, Martian surface, pockmarked with craters like those on the moon. This impression was confirmed between 1969 and 1971 as a flotilla of three other Mariner spacecraft

FIGURE 10.3   The Martian surface in the region around the Viking 2 lander. Some of the bigger rocks in the foreground are about 25 cm across. A soil sample from this area was analyzed by the Viking spacecraft life-detection instruments. No definitive evidence of Martian life was found. (Courtesy NASA)

scrutinized Mars, sending back detailed photographs and analyzing the Martian atmosphere. Instead of an abode for life as Lowell had imagined, the Mariner observations seemed to reveal that Mars was the cold, desertlike, dead world that Wallace had predicted.

Then in 1976, the two Viking orbiters and their landers provided even more detailed pictures of the Martian terrain and carried out a direct search for signs of life at two places on the surface (Figure 10.3). The Viking landers carried three life-detection experiments designed to look for traces of metabolic processes in the soil, as well as a gas chromatograph/mass spectrometer (GC/MS) that could search directly for organic compounds. Excitement was intense when two of the three experiments yielded results that initially suggested the presence of living organisms.

Subsequent analysis of these apparently positive life-detection experiments revealed that the results were probably due to unexpected Martian soil chemistry and not to the presence of life. The results could best be explained by the presence of a potent oxidizing agent in the Martian soils.

The Viking GC/MS instruments also searched for organic compounds and found none, even at the level of a few billionths of a gram per gram of Martian soil. But this level of sensitivity may not be good enough. A single bacterial cell weighs in at a mere trillionth of a gram, or 10^{-12} grams! About half of this is water, and the other half is organic compounds. Thus, if the Martian soils had in fact contained organic compounds equivalent to as many as about 10,000 bacterial cells per gram of soil, the Viking GC/MS system would not have detected them. We must do better than Viking to totally exclude the presence of microorganisms or even organic compounds in Martian samples.

A Habitable Early Mars?

After all this probing by the Viking spacecraft, was it finally safe to conclude, as Alfred Russel Wallace had predicted at the start of the twentieth century, that Mars really was "uninhabitable"? At first glance it might appear so, because it is possible using relatively simple reasoning to conclude that life could never have arisen there.

Over the last few years, at the behest of NASA, committees of distinguished scientists have been assembled by the National Academy of Sciences to define the region of the Solar System in which life could have arisen in the past or could still exist today. This region is known as the "habitable zone." The criteria are obvious—a source of energy, the presence of organic compounds, and, most importantly, temperatures compatible with the presence of liquid water.

The simplest way to define the habitable zone is to base it on the amount of solar energy received by a planet. Today, our Solar System's habitable zone can be estimated to extend from just inside the orbit of Earth to about halfway out to the orbit of Mars. In the distant past it would have been offset inward, because the Sun was less luminous. As the Sun continues to age and increases in luminosity, the habitable zone will shift outward.

If a planet has been within the habitable zone during the entire history of the Solar System, then it is said to be in the continuously habitable zone, or CHZ. Luckily for us, Earth falls within the CHZ. It is the only planet in the Solar System that does.

Early in its history, Venus may have been within the habitable zone. It is even possible that water was present on the Venusian surface. This means that at least some steps leading up to the origin of life could have taken place on Venus early in its history. But as the Sun grew hotter, the runaway greenhouse effect transformed the planet. Any organic remnants of early Venusian life must have been fried to gases long ago. We will probably never know whether Venus once harbored life, unless it left behind heat-resistant traces—or unless it consists of the sort of metallic creatures imagined by H. G. Wells!

As the Sun ages and gets brighter over the next billion years, Earth will face the same hellish fate as Venus. Our atmosphere now has sufficient concentrations of greenhouse gases—about a thirtieth of a percent of carbon dioxide, for instance—to generate a comfortable global temperature. But the Sun will increase in luminosity as it ages, so if the present mix of greenhouse gases remains unchanged temperatures will become increasingly unbearable. Even if the concentrations of these gases are lowered, there will be little additional cooling, because they are so low already. And they cannot be lowered very much because plants need a minimum amount of carbon dioxide in the air to be able to photosynthesize.

One obvious way by which the Earth's surface could be cooled further is if the cloud cover increases, reflecting more of the Sun's energy. A billion years from now, our descendants (if any survive) may rarely or never glimpse the Sun. They will be forced to live their lives in a humid twilight world of endless, gloomy drizzle, rather like the conditions that Herschel had imagined for Venus.

As the heat inexorably increases, even this cooling mechanism will be defeated. But perhaps by then the descendants of our inventive species, or some other intelligent race that has replaced us, will have learned to air-condition the planet. More likely, they will have departed for greener pastures, leaving behind a roasting cloud-covered Earth that will blaze like Venus in the light of the strengthening Sun and that will serve as a brilliant morning and evening star for a comfortably warm and habitable Mars.

Mars today lies outside the habitable zone and apparently it always has. Early in the Solar System's history, when the Sun was dimmer, Mars would have been even further outside the habitable zone. Temperatures would have been colder than today's frigid average, far too cold to permit the existence of liquid water.

But does this rule out the possibility that life could have ever originated and evolved on Mars? Perhaps not. As we learn more about Mars, we must be careful not to be fooled by our early observations. We have to be ready to expect the unexpected. And unexpected information about Mars is pouring in.

At the same time that the Viking landers were finding no apparent trace of life, images sent back by the Viking orbiters were showing that the first Mariner pictures had given a false impression of the Martian surface. In fact there were huge, deep canyons and towering volcanic peaks on the noncratered parts of the planet, evidence that Mars had been geologically active in the distant past. The numerous impact craters seen during the early flybys are confined to the oldest parts of the Martian surface, and indeed even that older surface has been much modified.

The possibility of an early Mars that was more hospitable to life than it is today has been reinforced recently by the Mars Global Surveyor spacecraft and the Pathfinder lander with its rover Sojourner, which arrived at Mars in 1997. High-resolution images taken from orbit by Surveyor, coupled with those taken directly on the surface by the Pathfinder, show that Mars has indeed had a rich and complex geologic history.

The pictures reveal a planet on which huge floods of both volcanic lava and groundwater surged across the surface in ancient times, the last flood perhaps only two billion years ago. Bright dunes, possibly made up of the mineral gypsum deposited by water flows, have piled up around the perimeters of Martian craters. At the Pathfinder landing site, rounded boulders litter the surface, apparently deposited when floodwaters raced through the region.

For the most part, the water sources were localized and transient, probably the result of a massive outflow of hidden groundwater that suddenly burst out onto the surface. Whether all these water flows were episodic, short-lived discharges or whether some were continuously flowing rivers that could have generated lakes or oceans is uncertain. Sometimes there is evidence that a sustained flow of water continued for a substantial period). The small channel on the upper right of Figure 10.4, about 200 meters across, is free of the dunes and collapsed material that are seen further down the valley.

Mars seems to be a planet shaped by a series of catastrophes. The Surveyor has shown clearly that nowhere on Mars are there broad valleys flanked by gently sloping hills, of the kind that are so common on Earth and that led James Hutton to propose his gradualist view of Earth's geology. But formations like the one in Figure 10.4 show clearly that it was apparently warm enough on Mars at one time for liquid

FIGURE 10.4   Part of a meandering canyon, which is about 2.5 kilometers wide, cutting through the Xanthe Terra region of Mars. The floor of the canyon on the upper right shows signs that slumps and landslides, wind, and sustained groundwater flow all contributed to its formation. (Courtesy NASA)

water to flow. So at least the first requirement for the origin of life appears to have been met.

In order for liquid water to have existed on the surface of early Mars for any length of time, temperatures must have been significantly warmer than today's bone-chilling average temperature of –58°C. The Martian atmosphere then was probably much denser than it is now, and this would have contributed to a greenhouse effect. But as James Kasting of Pennsylvania State University and his co-workers have shown, the Sun was so dim and Mars so chilly that most of the carbon dioxide in the atmosphere would have condensed out into dry ice snow or clouds of particles. Even adding copious quantities of methane would not have done the trick—Mars is so far from the Sun that there was simply not enough solar energy to warm the planet up. Something else is required to explain the liquid water.

François Forget of the Université Pierre et Marie Curie and Raymond Pierrehumbert of the University of Chicago have pointed out that clouds made up of dry ice particles have different properties from clouds of water droplets. Dry ice clouds will tend to trap outgoing heat radiating from the planet and reflect it back, like the panes of glass in a greenhouse. So a combination of high carbon dioxide levels and the presence of heat-trapping carbon dioxide clouds might have been enough to do the trick.

Today, however, even though the Martian atmosphere is still made up primarily of carbon dioxide, it is a mere wisp of gas compared with the thicker atmosphere of Earth and the hot, soupy atmosphere of Venus. What happened to the extra carbon dioxide that would have been needed to make early Mars a habitable world?

On Earth, as we saw in Chapter 3, carbon dioxide is removed from the atmosphere by the sedimentation of carbonate minerals in the ocean. Some of it is released again as the minerals metamorphose through heat and pressure, and some of this released gas in turn escapes into the atmosphere once more through volcanic activity, triggered in part by plate tectonics. Take away tectonic carbon recycling, and a growing fraction of the world's carbon would become buried as carbonates in ocean sediments. The resulting shift would turn Earth into a frozen world.

Unlike Earth, Mars is not now a tectonically active planet. If this were always the case, then carbon that had become sedimented as carbonate minerals on Mars could never have been tectonically recycled as it is on Earth.

However, there are other recycling mechanisms. During the planet's formation, internal radioactivity would have provided a heat source to partially melt the Martian interior and produce volcanism. Pictures taken by the Mars Orbital Camera of the Valles Marineris, a 4,000-kilometer-long canyon, show layers along the sides of the canyon that were probably formed by massive volcanic lava floods that took place during and even after the period of heavy bombardment ended (Figure 10.5).

Another way recycling could have taken place would have been through the heating and upheaval caused by objects from space slamming onto the Martian surface. Mars was heavily bombarded during the first several hundred million years of its history, just as Earth and its moon were during the same period.

On top of all this, it now appears that there may have been at least some plate tectonic activity on early Mars. Stripes of residual magnetism

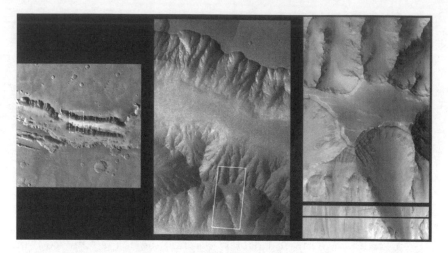

FIGURE 10.5   The Valles Marineris, a huge canyon system that stretches across the equatorial region of Mars. The left and center images were taken by the Viking orbiters. The outlines in these two images show the location of the higher-resolution right-hand image taken by the Mars Orbital Camera (MOC). The MOC image shows a 9.8-by-17.3-kilometer (6.1-by-10.7-mile) region in the canyon network. The layered rocks seen in these images imply complex and extremely active early geologic processes on Mars. (Courtesy NASA)

that alternate in their polarity have recently been detected on some of Mars's oldest regions.

Similar stripes are found on Earth's seafloor. They mark off regions where the seafloor has spread as a result of the upwelling of magma. Each stripe on the seafloor of our own planet signals a time when the Earth's north and south magnetic poles have switched places. The stripes on Mars are broader, suggesting that when Mars was young, its magnetic field switched north and south poles less frequently. There is little left of that magnetic field today, but this new finding suggests, astonishingly, that Mars may have had not only a liquid mantle but an ocean as well, and that the floor of the ocean was spreading just as ocean floors currently do on our planet.

If this did happen, then tectonics helped carbon to be recycled on Mars in the same way as on Earth. But on Mars, the recycling soon ground to a halt. The continued formation of carbonate minerals lowered atmospheric carbon dioxide levels, and temperatures dropped to well below the freezing point of water. Today the carbon that once was sufficiently plentiful in the Martian atmosphere to generate warm temperatures must be locked deep in the crust.

Although these various scenarios for the sequestration of carbonates seem plausible, there is a problem. So far, the direct analyses of Martian rocks by Pathfinder, as well as analyses of Martian meteorites, have shown that carbonates are not common. This is troubling, but the sample of rocks that have been examined is admittedly very small. One of the future goals for spacecraft exploration of Mars will be to look for those elusive deposits of carbonates. If they cannot be found, then the signs that liquid water was present during the early history of Mars become even more deeply puzzling.

As Martian surface temperatures cooled, what happened to the water? The amount of water present in the polar caps of Mars today is too small to have generated large bodies of standing water. Measurements by the Mars Orbiter Laser Altimeter indicate that if the ice caps of today were melted, they would produce a global water layer only about twenty meters deep.

This may seem like a lot, but Mars's polar caps have much less ice than the Greenland ice cap. It is estimated that to make a sea large enough to fill the basins on Mars would take at least ten times as much water. If there ever were extensive lakes or oceans on Mars in the past, that water has either been lost to space or is now hidden, possibly in the form of permafrost beneath the surface.

The question about whether there is still any water on Mars was supposed to be answered in December 1999 when the Mars Polar Lander descended for a soft landing near the south pole. In addition to a water detection instrument carried directly on the lander, during descent two basketball-sized probes were to be released which free-fell towards the surface. These two "Deep Space 2" probes, named Amundsen and Scott to honor the first explorers to reach the Earth's South Pole, were designed to slam into an area near the boundary of the southern polar ice cap at a velocity of nearly two hundred meters per second. Upon impact, bullet-shaped sections of the probes should have been released and penetrated about 1 meter into the surface. There, they would search for underground water using a miniature laser-based sensor. If a similar experience were carried out on the Earth's northern tundra, the probes would certainly reach the permafrost layer and detect the abundant water it contains.

Alas, something went very wrong. After radio contact with the spacecraft was automatically cut off as it started its descent toward the Martian surface, nothing was again heard from either the lander or the probes it was supposed to jettison. The fate of the mission remains a mystery al-

LIFE ELSEWHERE

though NASA has commissioned a team of experts to evaluate what may have happened and how to prevent similar failures in the future.

Although the loss of the Mars Polar Lander mission initially spread gloom through the planetary exploration community, the gloom was temporary. New missions are now being planned to search for Martian water and to reveal other aspects of the Red Planet's secrets, with launches scheduled for every two years starting in 2001.

What about organic compounds, the other component needed for the origin of life on Mars? Since it is not known whether Mars ever had a reducing atmosphere, it is difficult to evaluate the potential for home-grown organic synthesis on Mars. But certainly, throughout its history, Mars has received a steady input of objects from space. As with Earth, these have ranged from large bolides to tiny cosmic dust particles. Given that the Martian surface is very old and organic accumulation should have been taking place for a long time, Mars should in theory be dripping with organic compounds.

It is surprising, then, that no organic compounds were found by the Viking landers. Instead, there seem to be large amounts of a potent compound of an unknown type in the Martian soil. Any native organic material should have been destroyed very quickly by this oxidizing material. Of course, since the oxidizing layer may only extend a short way below the surface, organic material may still be present at greater depths.

As we start this new millennium, preparations are being made to probe and sample the Martian surface in more detail in a search for organic compounds. Spacecraft-based instruments to do this are currently being developed. One instrument, called the Mars Organic Detector, or MOD, is a small, compact device designed to carry out direct and highly sensitive amino acid and polycyclic aromatic hydrocarbon (PAHs) analyses of Martian samples. The first generation of MOD (Figure 10.6), which has been selected as part of the instrument package for the Mars 2003 lander, is designed only to examine samples of the Martian surface for the presence of very small quantities of amino acids, as well as PAHs. Later generations of MOD will determine whether any amino acids that are detected are left- or right-handed.

Far more complex missions are now being designed to return to Earth with a collection of Martian samples. These can then be analyzed by the best techniques we have available. If all goes according to plan, the first of these samples will arrive in April 2008. They can be examined for traces of fossil or even living organisms, something that cannot be done by the automatic probes on Mars's surface.

THE SPARK OF LIFE

Sample

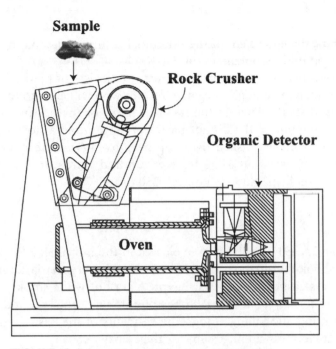

Rock Crusher

Organic Detector

Oven

FIGURE 10.6   A schematic diagram of the MOD instrument selected to be part of the 2003 Mars lander instrument package. A sample collected by a drill or scoop will be dropped into the rock crusher, which will first pulverize the sample and then drop it into the oven. After closing the oven at Mars ambient pressure (about a hundredth of that on Earth), the crushed sample will be step-wise heated to 950°C. Amino acids and polycyclic aromatic hydrocarbons (PAHs) in the sample will be vaporized (sublimed) and collected in an exposed area at the right end of the oven that is cooled to Martian night time temperatures (-105°C or less!). The released organic compounds will be detected using laser-based sensors. The entire instrument weighs about 2 kilograms and fits in your hand.

Even if all these analyses prove negative, the question of life on Mars will still not be settled. It is very difficult to prove a negative, and if traces of life are not found on the surface of Mars in spite of our most strenuous efforts, the question may still not be completely resolved until we drill into the Martian depths. Perhaps life has only managed to persist on this dry, cold planet at depths of hundreds or even thousands of meters, the equivalent of the microbial communities that have been discovered in deep boreholes on Earth.

There will surely be eternal optimists, regardless of how many samples are analyzed and give negative results, who will cling to the notion that life exists far down in the rocks of Mars, or that it existed on the planet sometime in the distant past. Carl Sagan told a story of an astronomer who was asked by a reporter to answer the question of whether life exists on Mars in five hundred words or less. The astronomer replied, "Nobody knows," 250 times.

Messengers from Mars

On June 28, 1911, at about nine o'clock in the morning, a thunderous boom shook the air above the oasis of El Nakhla el Baharia, about forty kilometers east of Alexandria, Egypt. A trail of white smoke was observed in the sky, and several stones plummeted to the ground, one of them hitting and killing a dog. The residents of the oasis assumed at first that God was stoning them for their sins!

Scientists from Cairo arrived on the scene a few days later. After inspecting some of the shiny black-skinned stones that had been dug out of farmers' fields, they determined that they were from a meteorite that had exploded as it entered the atmosphere. Some of the stones were sent to England and stored in the collections of the Natural History Museum in London (Figure 10.7).

During the 1980s the Nakhla stones, along with some other unusual meteorites that had similar textures and features, were found to contain gas bubbles identical in composition to the measurements Viking had made of the Martian atmosphere. Because no other place in the Solar System is known to have an atmospheric composition like that of Mars, it was concluded that these meteorites were fragments of the Martian crust that had been ejected into space by impact events.

In addition to the Nakhla stones, twelve other meteorites are now known to have come from the red planet. Around half have been recovered from Antarctic ice fields, where some have rested on the ice (and in puddles of water whenever the black stones were heated by the brief summer sun) for thousands of years since they fell from the sky.

When the comets and meteorites smashed into Mars, most of the bits of the Martian crust that were thrown up into the sky fell back on the planet itself. But a few reached escape velocity and went into orbit

FIGURE 10.7   One piece of Nakhla, a meteorite from Mars, from the
Natural History Museum (London). For scale, the cube is one centimeter on
a side. (Courtesy Johnson Space Center, NASA)

around the Sun. Most of these in turn eventually fell into the Sun or
into Jupiter, the two great gravitational vacuum cleaners of our Solar
System. A small percentage ended up in Earth-crossing orbits and eventually fell onto the Earth. In some cases, the transit time between
launch from Mars and infall to Earth is estimated to have been only a
few years, but the majority have taken millions of years to travel between the planets.

Could there be any evidence of possible Martian life in these meteorites? The problem of terrestrial contamination of meteorites poses an
almost insuperable difficulty. The methods now used for detecting
traces of organic material are so sophisticated that they can easily pick
up tiny amounts of material that the meteorites have acquired since they
landed on Earth. The Nakhla meteorites are a case in point. Trace quantities of amino acids apparently of terrestrial origin, and even structures
that look like bacteria, have recently been found deep inside the meteorites. It appears that either these were immediately introduced at the

LIFE ELSEWHERE

time the meteorites fell to Earth, or that cunning terrestrial bacteria had managed to work their way along fractures into the very hearts of the meteorites during the decades that the stones lay on a museum shelf.

In 1989 a report appeared in the journal *Nature* claiming that organic material had been detected in another Martian meteorite, called EETA79001, that had been recovered from the Antarctic ice sheet. Other groups immediately tried to verify these claims, and failed. Any traces of organic compounds, they concluded, were probably the result of terrestrial contamination. Surprisingly, this controversy received little attention from the press.

Such cautionary tales were all forgotten in August 1996, when a report published in the journal *Science* claimed startling findings about another Martian meteorite, ALH84001, that had also been found on the surface of the Antarctic ice. The teams of researchers from Stanford, the Johnson Space Center, and other institutions announced that there were traces of organic compounds in this potato-sized meteorite, particularly polycyclic aromatic hydrocarbons (PAHs) of the type that had earlier been discovered in the Murchison meteorite (and in Stanley Miller's flasks).

But their report went further, asserting that the meteorite also contained concretions of carbonate minerals that could have formed from the action of water, and even tiny cellular structures that could be attributed to ancient Martian life. The report was provocatively titled "Search for Past Life on Mars: Possible Relic Biogenic Activity in Martian Meteorite ALH84001." News of its imminent appearance was spread around the world even before it was officially published in Science. NASA, which had sponsored part of the work, basked in a blaze of publicity.

This paper has generated heated argument. At the same time, it has stimulated a resurgence of both scientific and public interest in the possibility of life on Mars. Because of the massive news coverage the report received, more than 60 percent of Americans surveyed in a poll thought that, at long last, life had been proven to exist on Mars.

Most scientists were highly skeptical, however. They began their attack by pointing out that the carbonate minerals found in the meteorite were probably formed at temperatures far too high for life to exist. David McKay of the Johnson Space Center and others involved in the original analysis have countered with newer evidence that the minerals could have formed at low temperatures.

Next, the meteorite was demonstrated to have been contaminated by terrestrial material. It seems that organic compounds produced by Earth's bacteria and algae had found their way into the meteorite during those brief Antarctic summers, just as they invaded the hearts of the Nakhla and other Martian meteorites. The NASA group has riposted that not all the organic material is terrestrial—they argue that some PAHs may be of Martian origin. The jury is still out on this question.

Most devastatingly, further research has shown that the small, fossil-like structures seen in the meteorite are too small to be real fossils. They are so tiny that only a single ribosome, the structure that cells use to make proteins, would be able to fit inside them!

McKay and his group have countered many but not all of these arguments, and the controversy continues. Regardless of the final outcome, the *Science* report did show one thing very clearly. A century after Lowell's claims of life on Mars, the possibility of Martian life is still able to generate enormous excitement and controversy.

Swapping in the Solar System

It is estimated that at present, several Martian meteorites find their way to Earth each year. And it is not just pieces of Mars that pelt our planet. Lunar meteorites have also been identified through their similarities to rock samples that were returned by the Apollo astronauts. Surprisingly, the totals of lunar and Martian meteorites are about the same. One would think that because the moon is so much closer, it should supply more material to Earth than the much more distant Mars. The explanation is simple: Mars is a bigger target and gets hit more often by bolides large enough to hurl pieces of the planet into orbit.

There may be some meteorites from Venus and Mercury, although we do not know enough about the composition of these planets to identify them. They are likely to be much rarer, because they must be flung "uphill" against the Sun's enormous gravity to reach Earth's orbit.

What about the reverse process, the transfer of bits of Earth to other bodies in the Solar System? Such events are probably uncommon, because it requires an escape velocity of more than 11 kilometers per second to launch pieces of Earth into space, compared to 5 kilometers per second for Mars and 2.5 for the moon. It takes much more energy to blast pieces of Earth off into space.

Nonetheless, at least a few Earth rocks must have been launched into solar orbit by severe impacts. Some must have found their way to our nearby Solar System neighbors. And this transfer must have been far more common early in the history of the Solar System, since the frequency of impacts was so much greater than it is today. This means that during the period of planetary infancy, when the origin and early evolution of life were taking place on Earth and perhaps on other planets as well, swapping of material between the planets must have been quite common.

It is therefore only logical to ask the obvious question. Could life have begun on one planet in the habitable zone, hitched a ride on one of its jettisoned rocks, and inoculated other worlds? The old idea of panspermia has been revived!

Perhaps life began on Mars even before it could have started on Earth. Mars, being smaller, would not have been subjected to as many sterilizing impacts by giant bolides. Perhaps an early Martian meteorite that found its way to the primitive Earth held hitchhiker organisms that seeded the planet with life. This possibility was hyped by both scientists and the press at the time of the announcement of possible evidence of life in the Martian meteorite ALH84001. One of the members of the science team at the first news conference, Stanford University chemist Richard Zare, proclaimed, "We are possibly all Martians."

But the journey of any life-seeding meteorite through space to another habitable planet is a hazardous one. As we saw in Chapter 1, organisms hitchhiking through space must survive intense exposure to cosmic rays and ultraviolet radiation. Then there is the strong likelihood of heat sterilization during the meteorite's fiery passage through the atmosphere of the new planet. Nevertheless, as Brett Gladman of the University of Toronto and Joseph Burns at Cornell University have noted, organisms in the center of a meteorite might be protected. After the meteorite landed on Earth, weathering might have exposed its interior. This would have released the alien organisms, allowing them to populate our globe.

Currently we are concerned with emerging diseases like the Ebola virus. Do we also have to worry about a potential biohazard from organisms that can hitchhike from Mars to Earth on the few Martian meteorites that land on Earth each century? Perhaps, but any organisms from Mars are far more likely to be rock-eating bacteria of the kind we met in Chapter 8 than anything pathogenic to humans or to the other organisms with which we come into day-to-day contact. They would, if they survived the journey to Earth, soon retreat to extreme environments that resemble those of their home planet.

If life on Earth really did originate on Mars, then any such recent Martian invaders would be virtually undetectable. They would be based on DNA, RNA, and the same twenty amino acids as occur in life on Earth. Alternatively, if Mars life is very different from our own, it could likely not compete with the vast array of organisms that have already colonized virtually every available niche on Earth. And what could Mars organisms eat in order to grow and reproduce? Either way, any such alien Martian invasions would be very different from those envisioned by H. G. Wells, by Michael Crichton in *The Andromeda Strain*, and by flying-saucer enthusiasts.

In the meantime, bits of Earth have been landing on Mars and other bodies in the Solar System as well. Consequently, some of our very distant bacterial relatives are possibly now quite happily living on Mars, or perhaps even on an asteroid. They would have had to undergo remarkable evolutionary adaptations to these harsh extraterrestrial environments. How will they affect us—and how will we affect them—when we finally meet each other again?

Habitable Moons

On July 9, 1979, the *Voyager 2* spacecraft returned some surprising pictures of Jupiter's moon Europa. About the size of the Earth's moon, Europa has a whitish, virtually noncratered surface consisting of water ice. The surface that was revealed by the low-resolution pictures sent back by *Voyager* was not completely smooth, however. There were short, sharp creases, lumpy regions that looked like icebergs frozen together, and long, sinuous lines that seemed to mark the boundaries of great plates of ice.

Now the more recent and far more advanced Galileo probe is swooping in a series of complex orbits that bring it repeatedly close to Europa's surface, and the pictures that it has been sending back are astonishing. This moon is turning out to be one of the most interesting bodies in the Solar System.

There are giant ice floes on Europa's surface, and between the cliffs that form the ice floe boundaries the surface appears to be newer, as if the floes had moved apart and new liquid or slushy material had welled up from below (Figure 10.8). The new material is sometimes reddish, showing that the ice is filled with mysterious colored compounds. Once the first upwelling of new material is completed, a slower oozing of cleaner water produces bright white lines that stretch for hundreds of kilometers across the surface.

Tectonic movement of the crust far below the ice appears to be powering all this activity. One result is huge strike-slip faults in the ice on the surface, some of them at least as large as California's San Andreas Fault. And the pattern of some of the reddish material that has been

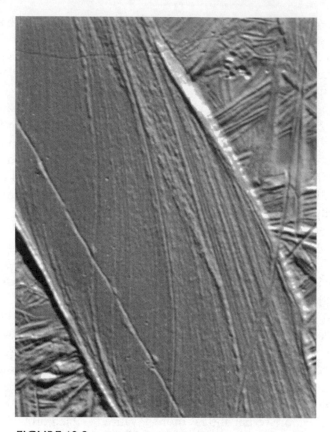

FIGURE 10.8   A Galileo spacecraft picture of the
south polar region of Jupiter's moon Europa. The gray
band is thought to have formed as plates on the icy
surface separated and material from below welled up
and filled in the widening gap. The curved, lined area is
a faultlike feature that extends over 800 kilometers (500
miles). It is roughly the same length as California's San
Andreas Fault, which stretches from the Mexican
border to San Francisco Bay. (Courtesy NASA)

spewed out onto the surface suggests that there are volcanoes underneath the ice, although no active volcanoes have yet been spotted.

Could this moon have a liquid ocean under its ice? And could there
be, on the floor of that ocean, the kinds of volcanoes and hydrothermal
vents that mark the floors of oceans on Earth?

One of the most convincing pieces of evidence for an ocean, and perhaps for tectonic activity, is that Galileo's instruments have found that Europa is surrounded by a weak magnetic field. One source for this could be a rotating body of salty fluid beneath the ice layer. This possibility has gained support from the recent discovery of hydrated salt deposits on the surface, in areas where the subsurface fluids have erupted onto the surface.

Europa has a density about midway between water ice and rock. It has a rocky core, on top of which lies a layer thought to consist of ice and water. As nearly as can be estimated, the ice layer may be ten to thirty kilometers thick and possibly a good deal thinner in certain spots. If the ice layer is thin enough to be disrupted periodically, as seems to be the case, then the ocean that lies below it must have an immense volume. It may be comparable in volume to Earth's oceans, even though Europa has only 6 percent of Earth's surface area. The liquid water ocean below the ice would then be a hundred kilometers deep. By comparison, the average depth of Earth's oceans is four kilometers.

And it appears that Europa is not the only moon of Jupiter with a subsurface ocean. The Galileo probe has found that Callisto, another Galilean satellite that orbits three times as far from Jupiter as Europa, also has a magnetic field, again apparently induced by a briny ocean far below a thick layer of ice.

How could liquid water possibly exist on these satellites, so far from the Sun and so far beyond the Solar System's presumed habitable zone? The answer lies in a combination of heat from radioactive elements deep inside the moons' rocky cores and heat generated from the repeated flexing of their interiors by the intense gravity of Jupiter. Even though the Galilean satellites always turn the same face toward Jupiter, their elliptical orbits mean that they swing a little bit back and forth on each orbit. Remarkably, the tidal friction that is induced by Jupiter's gravity during these short swings is enough to heat them up.

The tides on the moons are huge because Jupiter is so big and so close. As we explained in Chapter 3, after Earth's moon formed, tidal forces quickly caused it to move a long way away from Earth. And yet, even though Jupiter's gravitational pull is so much stronger than that of Earth, the Galilean satellites have not moved proportionately further away—Io, the closest of Jupiter's large satellites, moves in an orbit about as far from Jupiter as Earth's moon is from Earth.

The primary reason these satellites have stayed relatively close to their planet is that the moons of Jupiter tend to lock together—when one speeds up, the others tend to slow it down and vice versa. A second reason is that Jupiter is so massive. Because Earth spins faster than the moon and in the same direction, the wave of gravitation-induced distortion on Earth always moves ahead of the position of the moon in the sky. This wave pulls forward on the moon, speeding it up slightly and causing it to move further away. But since Jupiter's moons are tiny relative to Jupiter and deform the planet only slightly, this effect is much smaller in the Jovian system. The result of all these effects is that tidal friction heats up Jupiter's moons, even though they lie far beyond the habitable zone.

Organic materials, too, are probably present in abundance on Europa. Its atmosphere has long since been lost because of its weak gravity, but at the outset it probably had a reducing atmosphere like Titan's. Europa is so far from the Sun that the methane and ammonia in its atmosphere would not have been instantly destroyed by ultraviolet light. The infall of comets and of carbonaceous meteorites like Murchison could also have provided a rich source of the ingredients needed for prebiotic syntheses along with some intact organic molecules. In addition, there is a very large nearby source of the organic molecules, as well as raw materials for abiotic organic synthesis—Jupiter itself.

Jupiter is one vast Miller-Urey experiment that has been running, not for weeks, but for billions of years. It has methane, ammonia, and hydrogen cyanide in abundance, although there seems to be very little water in the upper atmosphere. And numerous giant lightning storms, the equivalent of Miller's spark generator on a vast scale, have been seen on the planet's night side. These were observed in detail by space probes that ventured beyond Jupiter's orbit and gave us for the first time a glimpse of the planet's night side. Regrettably, the only probe that has yet penetrated Jupiter's atmosphere, carried by the Galileo spacecraft, was not designed to detect organic materials.

On Jupiter itself, of course, the various steps required for the origin of life beyond the synthesis of simple organic compounds would probably have been impossible. The solid core of the planet lies far below the surface and is too hot to support life. Any living organisms would have been forced to evolve entirely as free-floating creatures, unlike Carl Sagan's hypothetical Venusians. Sagan had supposed that floating Venusians would only have evolved after many generations of living on the ground.

The spectacular Shoemaker-Levy comet impacts in 1994 showed vividly that Jupiter, with its powerful gravitational field, often pulls in comets and asteroids and that the impacts send vast plumes of material hurtling skyward. Similar impacts in the past could have carried substantial amounts of material from Jupiter's atmosphere to its moons, particularly during the hectic early period of the Solar System's formation, when the moons were far closer.

All these new discoveries and ideas have made the concept of a habitable zone based solely on the Sun's energy seem quaint indeed. There are more places in the Solar System where life could lurk than would have been imagined just a few years ago.

Life Beneath the Ice

Even though Europa's ocean is incredibly deep and dark, it is not impossible to imagine life surviving there. The pressure at the ocean bottom would not be utterly crushing, as it might be on Earth if our planet had an ocean a hundred kilometers deep. Because the gravity of Europa is only a seventh that of Earth, the pressure at the bottom of its ocean is only a little over a thousand Earth atmospheres. Life can accommodate itself to this—the bottom of the Marianas Trench, the deepest part of Earth's ocean, is at nearly the same pressure, and a shark has been spotted swimming about at this depth.

But if the ocean on Europa is not only deep and dark but utterly still, then it would not be a very likely place for life to begin. This is why the discovery of possible plate-tectonic-like movement beneath that ocean is so exciting. Are there undersea vents, and has life evolved at the vents? This would be the ultimate test for the ventist school of the origin of life that we discussed in Chapter 4. We cannot yet see below the ice, but we will soon learn much more about this moon from some of the imaginative probes being designed to visit Europa over the next few years.

On the drawing boards is an orbiter that will use radar to scan Europa's surface beneath its ocean. A lander is being contemplated that will measure seismic activity and ice movement. One ingenious project with a good chance for funding (Congress willing!) in the near future is called the Europa Ice Clipper. It is a probe designed to drop into low orbit around Europa. Once in position, it will fire a heavy object at the satellite's surface. During subsequent orbits it will scoop up some of the ice particles that explode upward from the impact site, and finally it will return this precious sample to Earth. If the Ice Clipper is lucky enough to gather some of the puzzling reddish material that lies on the recently resurfaced parts of Europa, we may gain important answers about life on this moon without having to penetrate all the way through the ice to its ocean.

At the very limit of modern technological feasibility is a Europa lander that would use a radioactively heated minisubmarine-like probe to tunnel down through the ice and into the ocean far below. The probe, if it is to send information back to the surface, will have to trail a long, thin wire or cable behind it as it melts its way down. It will be some time, we suspect, before such a project is funded, but although the technology seems daunting, there is now a wonderful opportunity to practice it here on Earth.

During the 1970s, radar surveys of Antarctica revealed about seventy freshwater lakes that are trapped like liquid bubbles beneath the ice. These lakes are apparently kept liquid by heat diffusing up from the crust below. Many of them are covered by layers of ice that are kilometers thick.

The largest of these hidden bodies of water is Lake Vostok, named after a Russian Antarctic Research station fortuitously situated on the ice cap directly above it. Even though the lake is the size of Lake Ontario, there is little evidence of it on the surface, for it is overlaid by more than four kilometers of ice.

The Russians have been drilling down toward this lake since 1974, extracting with exquisite care an ice core that has given a record of almost half a million years of the Earth's climatic history. The true extent of the lake was, however, only revealed in 1996, just as the borehole approached within about a hundred meters. Drilling was suspended during that winter and halted entirely when the scientists realized that breaking through into the pristine lake might contaminate it irreversibly. Further, since the lake is under very high pressure, water from it could have squirted up and created a dangerous high-pressure geyser on the surface.

The U.S. National Science Foundation has now begun a program to penetrate that last hundred meters and explore the dark and hidden lake. In addition to the marvelous scientific opportunity that the lake provides, there will also be an opportunity to test devices of the kind that can be used to penetrate to Europa's ocean. The trick will be to obtain pristine samples from the lake, which has been isolated from the surface for at least a million years. What organisms will be found there? Has their DNA diverged from that of their relatives on the surface? Will it be possible to penetrate all the way through the lake to the fifty meters of sediment that have accumulated on the bottom? And what organisms will the probes find, living and fossilized, in the sediment?

After we practice these techniques on Earth, we will be ready for Europa. This remarkable moon is so exciting that it is sure to be explored in detail in the future. There may, of course, be no life there. But even if

Europa and the other giant moons of the outer Solar System are lifeless at the present time, there is hope for them in the future. Titan may easily turn out to have a rich mixture of prebiotic organic compounds, even though at present it is much too cold for life as we know it. Moreover, Europa and the other Galilean moons Ganymede and Callisto may also be worlds on which a frozen or ice-covered prebiotic soup never had a chance to go through the additional processes that would have led to primitive self-replicating molecules. But the Sun will continue to increase in luminosity as it ages, warming up these moons like giant frozen casseroles. After a delay of billions of years, life could eventually appear on them spontaneously—provided that we do not first contaminate them with terrestrial organisms. (For more on this problem of contamination, see note to p. 296.)

As we begin to explore our own and other solar systems, we and our descendents will undoubtedly find planets on which life has come and gone. We will also find many others on which life has the potential to appear in the far distant future.

Life Beyond the Solar System

Barnard's Star is like a frisky colt in a heavenly pasture full of staid adult horses. It is about six light-years away and is moving in our direction at astonishing speed, covering a light-year every five thousand years. It is also moving swiftly across the field of more distant stars, so that pictures taken of it even days apart can detect its motion.

This nearby star consumed the working life of astronomer Peter van de Kamp of Swarthmore University. From 1938 to 1962, he took thousands of plates of its position in the sky. Plotting out the position of the star on each plate, he detected what seemed to be a slight wobble as it moved across the star field. He calculated that the wobble must be due to the presence of a large planet, $1^1/_2$ times the size of Jupiter, moving in orbit around the star once every twenty-four years.

If you pick up a tiny child and swing it around, the child traces a large circle and your shoulders trace a smaller one. Van de Kamp's astrometric calculations were based on this phenomenon. When a large planet orbits its star, both the planet and the star actually orbit around a point lying between them. Because the star is so much more massive, the point lies very close to the star; indeed, it may actually lie beneath its surface. Even though the planet is so faint as to be invisible, the star's circular motion can be followed by the wobbles in its position relative to other more distant stars in the same telescope field of view.

If van de Kamp was right, the Barnard's Star planet would have been the first planet to have been found outside our own Solar System. But alas, he wasn't right. The wobble was apparently due to problems with his telescope. More recent analyses have found no detectable wobble in Barnard's Star.

Right up to the 1980s, there were several other announcements of the discovery of extrasolar planets, including some apparently quite convincing evidence for planets orbiting another nearby star, Van Biesbroeck 8. But this discovery too turned out later to be incorrect.

During the last few years of the twentieth century, however, these early failures have been replaced by a torrent of successes, spurred by increased sensitivity of astrometric and spectrophotometric measurements and the use of sophisticated computer programs that can subtract the motion of the Earth and our Sun. Extrasolar planets, at least large ones, turn out to be surprisingly abundant.

Most extrasolar planets are impossible to see directly with telescopes (although see Figure 3.1 for one startling possible exception). Even Jupiter-sized planets are so faint that from great distances, they are obscured by the glaring light of the host star. So, instead of seeing the planets firsthand, astronomers have used two main methods and a number of clever minor ones to detect them.

The first method is based on the astrometric technique, which directly measures a star's wobble. These measurements can now be carried out with better telescopes than the one van de Kamp used. The second method is more sensitive and depends on the Doppler effect, the small shift in the wavelength of the star's light that an observer perceives as the star moves toward and away from Earth. The frequency of light that we measure from the star changes as it wobbles back and forth, just as the pitch of the sound of a train whistle appears to change as the train races past an observer.

By measuring this minute overall motion of the star and using Newton's laws of motion, the sizes and rotation times of any planets near the star can be calculated. Only large Jupiter-sized planets can be detected, however, because the tiny wobble in a star produced by smaller, earthlike planets cannot yet be measured. Furthermore, it is easier to detect a planet that is close to its star and moving quickly than one further away and moving more slowly. When wobbles or changes in light frequency are spread out over a span of decades, they can be very hard to separate from background measurement fluctuations.

Even given these constraints, a growing number of teams of astronomers have been finding telltale wobbles in many nearby stars. The most successful have been planet-finders extraordinaire Geoffrey Marcy, at San Francisco State University, and Paul Butler, now at the Carnegie Institute of Washington. Between the fall of 1995 and June 1999, these and other astronomers have found twenty-one planets orbiting nineteen different stars of various types, with exotic names like 51 Pegasi, 55 Cancri, and 16 Cygni B. In most cases, only the largest planet in each of these solar systems has been found. But in April 1999, the Marcy-Butler team, along with scientists from the Harvard-Smithsonian Center for Astrophysics, announced that they had discovered a family of three planets orbiting the star Upsilon Andromedae.

All these discoveries appear to be of relatively ordinary solar systems. But far more outré systems have also been discovered. Two are planetary systems found circling around pulsars.

Pulsars are incredibly dense neutron-star remnants of supernova explosions. Like madly spinning lighthouses, they rotate at high speed and send out intense beams of radiation. Their planets must be unlike anything we know, for the supernova explosions probably destroyed any planets originally associated with the exploded star. The pulsar planets most likely formed from bits of debris orbiting around the remnant pulsar, and must be extremely radioactive as a result. On top of this, the planets would be bathed in such intense blasts of radiation from their angry, whirling host pulsars that life on them would be unimaginable.

In sum, the good news is that planets around other stars have finally been found. The bad news is that most of these other solar systems do not seem to resemble ours. The planets that have been found are generally huge, much larger than Jupiter. They may be even larger than we think, because the masses that have been calculated are minimum estimates. In many cases, we do not know the plane of the ecliptic of these new solar systems. If the plane is tilted away from us, the stellar wobbles will actually be larger than our measurements indicate, and the planets will be correspondingly more massive. Some of these new planets might actually be brown dwarfs, objects so massive that they have almost become stars themselves.

Contrast the extrasolar planets that have been found so far with what an observer on some distant world would see looking at our Sun. Very good instruments might detect the motion caused by two massive planets (Jupiter and Saturn) orbiting far out from the Sun. This makes sense from the standpoint of models that assume that the lighter materials had been driven out to the perimeter of the disk during the Solar System's formation and had then condensed into gas giant planets. The heavier materials would have been left behind to form rocky planets like Earth.

Nevertheless, solar systems like ours seem to be a rarity. The stars 16 Cygni B and 47 Ursae Majoris have planets with masses at least two times that of Jupiter in Mars-like orbits, so that they do somewhat resemble our own Solar System. But the planet orbiting Tau Bootis, estimated to be about four times as big as Jupiter, is so close to its star that it makes one complete revolution around it in only 3.3 days! It is much closer to its star than Mercury is to ours, for Mercury completes its journey around our Sun in a relatively leisurely 88 days.

The Tau Bootis planet is not unusual. A massive planet at least half the size of Jupiter makes a complete trip around 51 Pegasi in just 4-1/2 days. Most other extrasolar planets that have been discovered also crowd close in to their stars and orbit around them at breakneck speed.

Other solar systems seem to have a mixture of features. The system around Upsilon Andromedae consists of three planets: a very close planet like that of 51 Pegasi, a planet twice the mass of Jupiter that revolves at a distance a little less than that of Earth, and an even more massive planet, the size of whose orbit lies between that of Mars and Jupiter.

We suspect that as extrasolar planets continue to be discovered, especially ones that take many years of observation to be detected, these first solar systems to be discovered will turn out to be unusual and most solar systems will be found to resemble our own. By the time you read this book, many more extrasolar planets will have been discovered. As we saw in Chapter 3, even the formation of planets (see Plate 3) can now be detected. Planet formation appears to be a very robust process, and stars that have planetary companions are likely to be common rather than rare.

Marcy and Butler estimate that 10 percent of all stars in our Milky Way Galaxy have planets of one sort or another. That means that in our galaxy alone, there are roughly 10 billion solar systems. If each solar system has several planets—ours has nine—then the number of planets in the Milky Way must be close to the number of stars! Surely there are earthlike planets out there, probably many of them.

Now that we know something about the extrasolar planets that have been discovered so far, we can speculate about the potential for life on them. If we apply the conventional habitable zone criteria, in most cases the prospects look pretty dim. It might be supposed that the surface of a 51 Pegasi–like planet, for example, must be so scorching hot that any gaseous components and water would have long ago evaporated into space. Yet new data from light reductions when such planets pass between us and their star show that these planets really are so large that they must be gas giants. The more we learn about these extrasolar planets, the more the mystery about them deepens.

If some of these massive planets turn out to be rocky bodies rather than gas giants, then if they are far enough from their suns they may have liquid water. It is quite possible that all the steps that gave rise to the origin of life on Earth have taken place on these extrasolar worlds and that they are teeming with life. Presumably, such life would be short, squat, and muscular, for the gravitational fields of these giant planets will be very powerful!

Whether or not the planets themselves are gas giants, they will probably have extensive systems of moons. And since the planets are larger than Jupiter, some of their moons would likely be huge as well.

What would an Earth-sized moon revolving around the outermost of the three known planets of the Upsilon Andromedae system be like? Although the planet is further from its star than Earth is from the Sun, its star is substantially brighter than our Sun. As a result, the differences probably average out. The history of the planet's large moon would probably be much like that of Earth. Heat generated from the decay of radioactive elements would have melted the moon's interior, releasing water and other gases. These volatiles would have been vented to the surface by volcanic eruptions, where they would have formed an atmosphere.

Greenhouse gases in the atmosphere would generate temperatures allowing liquid water to exist, so that oceans or lakes would be present on the moon's surface, not buried under kilometers of ice like on Europa. Lightning flashes in the atmosphere and ultraviolet light would drive prebiotic syntheses. A prebiotic soup could have formed that contained both moon-grown components and those derived from debris infalling from space. Given all this, at some point a space-faring Darwin would surely have turned up to start the processes leading to the emergence of life.

But even if all of these events have taken place on this extrasolar planet-sized moon, the environment there would still be dramatically different from that on Earth. The giant host planet will exert a massive gravitational pull on the moon, generating huge tides in its oceans. The moon will always have one face turned toward its planet, but the oceans will slosh violently back and forth as the moon wiggles slightly in its planet's grip. There will likely be massive volcanic activity driven by the tidal heating. If the moon is in a tight orbit, the planet will loom ominously in its sky, causing prolonged eclipses of the Sun. If it is far from its planet, like Jupiter's Callisto, the moon could swing as much as 2 or 3 million kilometers toward and away from its star every two or three weeks. If on top of this the moon's planet has an elliptical orbit like that of Earth, the moon would move 10 or 12 million kilometers toward and away from its parent star over the period of a year, in a very complex, chaotic pattern that would rarely repeat itself.

If the axis of the moon is tilted, the seasons that result would be greatly complicated by all these orbital movements. Temperatures would tend to fluctuate dramatically and unpredictably, perhaps causing the oceans to freeze over and thaw again on a very short timescale. Summers and winters would cycle over the two- or three-week orbital period, and each would be unpredictably mild or extreme, depending on where in their orbits the moon and its planet happened to be.

Of course, a thick atmosphere and large oceans like those of the Earth would tend to damp out these cycles, just as the oceans moderate the climate and buffer the severe effects of winter on Earth. But if the moon has a thin atmosphere and small seas, the cyclic events on its surface would be even more extreme than those that took place on primitive Earth. Such extreme conditions would have driven proto-Darwinian selection on the molecules in its oceans even more effectively than on Earth.

If the ideas we have suggested in this book are correct, then the likelihood of life arising on such a moon-world would be very high. And if life did arise on this moon-world, what have the forces of evolution yielded, as its inhabitants have been forced to wrestle with such a crazily fluctuating environment?

Life on Rogue Planets!

Very early in the history of our Solar System, many more planets were circling around the young Sun than there are today. It was one of these planetary bodies that hit Earth and created our moon. Others, because their orbits were unstable, were captured and devoured by the young

Sun. Still others collided and merged into bigger planets. But some, and perhaps the majority, came too close to Jupiter or Saturn and were flung out of the Solar System altogether into the dark emptiness of interstellar space.

Our own Solar System was not the only one hurling out such rogue planets. As we saw in Chapter 3, a binary star system in the Taurus constellation may have been caught in the act of jettisoning a Jupiter-like planet. If all solar systems did so during their formation, interstellar space should be filled with such dark objects. And perhaps these planets are not all as dark and lonely as the dying world that Byron envisioned when he imagined that the Sun had been suddenly quenched.

David Stevenson of the California Institute of Technology suggested in 1999 that these rogue interstellar planets might also harbor life. Because these planets were ejected from their parent solar systems very early, they should be rich in volatile materials. Moreover, their initial atmospheres would not have been lost through the escape of light gases like hydrogen, nor would they have been blasted away by intense solar radiation or periodic impacts of large chunks of debris left over after planetary accretion.

Of course there would be no sunlight to illuminate and warm these worlds floating about in the vastness of space. However, if they are earthlike in size, then the natural radioactivity in the rocks would be sufficient to melt their interiors. This would liberate water and other volatile components. Because of their dense atmospheres, consisting primarily of hydrogen, heat leaking out of the planets' interiors would not have been lost rapidly into space. As a result, surface temperatures could be above the freezing point of water. Water vented to the surface could condense, and oceans could form. All of this in the void of space, surrounded by temperatures close to absolute zero!

The potential for prebiotic syntheses and chemical evolution will last for billions of years on these interstellar worlds before the heat-providing radioactive elements decay away. The origin of life on Earth took place over timescales of less than several hundred million years and perhaps, according to some, in a period shorter than 10 million years. There is no obvious reason why life should not originate on these rogue planets as well.

On these bizarre, dark worlds, what would the forces of evolution produce? Obviously, life could not gain its energy from photosynthesis, so it might never progress beyond the unicellular stage. But evolution is very inventive; it may be that multicellular organisms that gain their energy from the abundant hydrogen have evolved or will evolve on these worlds.

Planets in the dark emptiness of interstellar space teeming with life! It all sounds like something out of a *Star Trek* episode. But according to Stevenson, there might be so many such planets that they could easily be the commonest places for life in the universe.

How could we find these life-sustaining rogue worlds? They would be invisible except when they occasionally pass between us and a distant star. But unfortunately, objects in the Oort cloud of comets that populates the outer reaches of our Solar System can briefly block stars as well.

Unlike the Oort objects, however, these planets should radiate substantial amounts of heat, which means that they might be detectable in the infrared. The new space infrared telescope facility, the last of NASA's large observatories, is scheduled to be launched in 2001, and it might find some of these rogue planets that happen to be passing near us. We should be able to distinguish them from more distant objects by their swift motion across the sky. The 1999 sighting of an apparent large planet (or small brown dwarf) in interstellar space by a telescope in Chile suggests that there are many smaller, cooler planets between the stars.

Perhaps if some of these planets have developed civilizations, they will find us rather than vice versa. But they may not choose to, for such intelligent beings would not enjoy visits to unpleasant planets such as Earth that are bathed in nasty actinic radiation from their nearby stars and surrounded by atmospheres loaded with corrosive and poisonous oxygen.

Can We Recognize Life if We Find It?

In December of 1990, a year after its launch, the Galileo spacecraft had circled the Sun and returned almost to its starting point. It passed within 960 kilometers of Earth, which gave it the final burst of speed it needed to reach Jupiter. As it hurtled by its home planet, it made a series of observations using its onboard instruments. The spacecraft detected abundant oxygen in Earth's atmosphere, along with evidence of liquid water. Pictures taken with its onboard camera showed extensive land areas that had some sort of pigment on them. In addition, narrowband, pulsed radio transmissions were detected. Carl Sagan and the other scientists who conducted this study of Earth from space concluded that taken together, this evidence implied that not just life, but intelligent life, existed on Earth.

Of course we knew that already, although we are sometimes not sure about the intelligent part. But it is comforting to discover that if life has appeared elsewhere, has developed a capacity for photosynthesis, and has evolved to the point of inventing technological abilities such as radio transmissions, we should be able to detect it using the kinds of instruments currently on our space probes.

But there are other problems with the detection of life. If Earth is any example, intelligent life and even photosynthetic life only emerges after a long history of evolution. We do not want to miss detecting less advanced life, or even oceans full of prebiotic soup, as we search for life elsewhere.

Consider what a spacecraft like Galileo would have observed if it had swung by Earth three billion years ago. By that time, life had developed to the point that photosynthesis had been invented, but no significant amounts of oxygen had yet accumulated in the atmosphere. Any life on the land surfaces was confined to the intertidal zones and consisted of little more than layers of bacteria. These were heaped up into structures called stromatolites that huddled in little hummocks under the intense ultraviolet radiation. And of course there were no humans to produce radio signals. Would the spacecraft have detected the presence of life, even if it had passed very close to Earth?

We have a good chance of finding primitive life, if it exists, in our own Solar System because we have the capability to send spacecraft to explore and sample the surfaces of promising planets and their moons. But if we are to extend the search to other solar systems around other stars, we will probably be forced to look for much more advanced life, more like our own.

Finding Life on Planets Beyond the Solar System

Astronauts circling above Earth's atmosphere are awestruck by the velvety black of space. But space, at least in the vicinity of Earth, is not as black as it seems.

The reason for this is that we are surrounded by a virtually invisible cloud of dust particles, left over from the Solar System's formation. You can actually see part of this cloud, known as the *gegenschein*, on a very dark and clear night, provided that you are far from city lights. It appears as a diffuse brightening in the night sky, directly opposite the position of the Sun below the horizon. The gegenschein is caused by the reflection of light from these tiny particles as they are illuminated by the Sun behind us.

Like the tiny particles that Tyndall found in the air, the particles of the inner Solar System are omnipresent in the space around us, and together they form the zodiacal light. The gegenschein is the manifestation of the zodiacal light that we can see most easily. But the light is everywhere, and it can interfere with the ability of space telescopes to see very faint objects.

The great problem that we face as we try to observe planets or moons the size of Earth, even those that are circling nearby stars, is that they are incredibly faint. Even mighty Jupiter is only a billionth as bright as the Sun, and seen from twenty or thirty light-years away it would be far fainter than the faintest galaxies currently observable by the Hubble Space Telescope. To see planets that are even smaller and fainter, we must put our telescopes beyond the zodiacal light. The only way to do this is to send them beyond the orbit of Jupiter, where the cloud of dust thins out to practically nothing. Only there will the sky be sufficiently black.

The telescopes must be specially designed, for the blinding light from a star will overwhelm the light from its planets, even under the most ideal seeing conditions. One way to overcome this is to use a technique called interferometry. A series of coordinated telescopes, working in tandem and rigidly connected to each other, can use light wave interference methods to cancel out the light of a star. Planets near the star will shine with light that is out of phase with that of the star, and should then be detectable by the telescopes.

Already, present-day technology allows the light from stars to be canceled out in this way. One design for such a space telescope, currently on the drawing boards, consists of four 1-meter telescopes arranged along a rigid 75-meter beam.

Unfortunately, such a telescope would be unavoidably massive. Transporting it beyond Jupiter would be a very expensive proposition, with an estimated price tag in the $500 million range. Although there is as yet no funding for such an instrument, Roger Angel and Neville Woolf at the University of Arizona, along with Ronald Bracewell at Stanford, have pioneered the developmental concepts of the interferometry method. They have shown that the results would be worth it. Such a telescope, looking at our Sun from a nearby star, would be able to detect most of the planets in our Solar System, including Earth, and could determine the chemical composition of their atmospheres.

As Daniel Goldin realized, nothing more exciting is likely to emerge from the space program than the discovery of a planet with an earthlike atmosphere in some other solar system. Unequivocal evidence for life beyond our own little speck in the galaxy—and evidence for complex, highly developed life—would catapult our exploration of space into high gear.

There is absolutely no doubt that such extrasolar planets will be found in the lifetimes of most people reading this book. Can our species bridge the immense distances between stars in order to reach them? It is not obvious at the moment how this can be done, unless we develop *Star Trek* warp-speed technology. The distances are daunting.

Pioneer 10, launched in March 1972 and the first spacecraft to leave the confines of our Solar System, is only a thousandth of a light-year (about seven billion miles) from Earth, and it is continuing to slow down as the distant Sun tugs at it. By comparison, the 47 Ursae Majoris solar system with its Jupiter-like planet is fifty light-years away. The universe is an immense place.

We are confident, however, that the discovery of life-bearing extrasolar planets will stimulate some very clever thinking about how to visit them. Our species has come a long way since it first emerged five million years ago in the forests of eastern Africa, and the pace of our exploration and invention is accelerating at an exponential rate. Like no other species on our planet, we are driven by a need to know. And that drive is sure to carry us out into the galaxy to meet the other curiosity-driven creatures that live there.

What Would the Ultimate Discovery Tell Us?

Since this book is about the origin of life, the discovery of life elsewhere is an important part of that story. But as we catch our breath after this dizzying tour of the universe, we must remember one thing. Even when we do discover life on other planets, we may be no closer to finding out how life began.

If the extraterrestrial life that we find turns out to be biochemically and genetically distinct from life on our own planet, it will show us that living organisms could have arisen in more than one way from a primordial soup. But even if the life that we discover is only unicellular, it will probably be highly evolved, just as the bacteria and archaea of Earth are. This means that the origin of these extraterrestrial bacteria is likely to be just as puzzling as the origin of life on our own planet.

EPILOGUE

A speculation which is, within a few centuries, susceptible of observational disproof may perhaps claim to be described as a scientific hypothesis.

J.B.S. *Haldane*, American Scientist *33 (1945): 145*

THE SEARCH FOR LIFE ON OTHER PLANETS IS ACCELERATING, and there are likely to be incredibly exciting breakthroughs very soon. But it is no substitute for the search, here on Earth, for that elusive Golden Spike. To really understand the origin of life, we must determine the steps needed to create it in the laboratory. Only then will we have answered the questions that were posed at the start of this book.

How far are we from the time when those two groups of scientists meet each other, quaff a few brews, and finally hammer in the Golden Spike? There are quite a few barren hills to climb, chasms to cross, and unexpected rushing torrents to bridge before we get there. But a few scouts have been sent ahead to scan the terrain, and we can now begin to see the outline of what still has to be accomplished.

Is It Truly Possible to Create Life in the Laboratory?

The scientific community used to think that because the origin of life must have been a very improbable event, a considerable length of time was required for all the steps to happen in the right sequence. In the 1950s, paleontologist George Gaylord Simpson used evidence from the then-known fossil record to suggest that life only began about a billion years ago and that it must therefore have taken over two billion years to arise. Such a timescale certainly puts the process of life formation far outside anything that we could imagine replicating in the laboratory. Luckily, we now know from both fossil and molecular evidence that Simpson's estimate was wildly wrong. Even so, the span of time during which life originated might still be too vast for us to contemplate doing it all ourselves.

Bill Schopf has found definitive fossil traces of the presence of life in 3.5-billion-year-old rocks, and there is some less certain evidence that life was present 3.8 billion years ago. Some of the genealogies of the genes that we explored in Chapter 9 have been traced back approximately as far. This means that life probably arose between about 4 billion and 3.8 billion years ago. Thus, the span of time between the last of the great sterilizing asteroid impacts and the origin of life is probably about 200 million years.

Antonio Lazcano of the University of Mexico and Stanley Miller feel that the origin of life could not have been stretched out over such relatively long spans of time. They point out that the primordial soup would have been destroyed by circulation of the ocean through hydrothermal vents in a mere 10 million years. The origin of life must have taken less time than that, particularly if it had to happen while the primordial soup was still rich and nourishing.

It also seems unlikely that chemical evolution took hundreds of millions of years, for most of the key organic chemicals could not have persisted for so long. Our own hunch is that even 10 million years is probably a substantial overestimate. Perhaps life appeared over a span of a mere few thousand years. We hope and expect that it took even less, for otherwise the attempts by scientists to produce life in the laboratory are doomed to failure. Granting agencies provide money for periods of from two to five years, not millions of years or even thousands!

In the absence of such long-running grants, and an infinite supply of graduate students, what can be done in the laboratory? If immense amounts of time really are required, then the task would be hopeless. Thus, there has been a powerful urge among scientists to try to short-circuit the process—to cut to the chase, so to speak.

This is why there has been so much interest in producing a self-replicating molecule in the laboratory. The discovery of such a molecule would seem to solve the problem immediately—the lucky scientist who finds it would be able to cry, "Shazam! I have created life!"

But we have seen that it is unlikely that a truly self-replicating molecule will be found. The first genes almost certainly evolved in concert with other "helper" molecules that aided their replication. So if we cannot cut to the chase, what can we do?

There is still hope. Prebiotic chemistry experiments have shown repeatedly that many quite remarkable chemical reactions, once thought to be only the province of living organisms, can take place in the absence of enzymes. The experiments with PNA, and the laboratory selection experiments that have begun to explore the RNA world, have given us a glimpse of how genes might have evolved. All of this work has shown just how complicated and varied the world of prebiotic chemistry must have been.

These discoveries will do us little good until we can find how these molecules can be sorted out and combined with each other, so that more complex and integrated molecular aggregates can form. We must enlist the aid of Darwin, or rather of the proto-Darwin whom we met in Chapter 5.

This can only be done by creating, in the laboratory, imitations of the early Earth. These laboratory experiments will have to be designed to mimic the variety of conditions that must have been present at that distant time. Very little has yet been done using this approach, and we do not know how far it can be taken in the laboratory. But we have not yet harnessed the full power of proto-Darwinism, in which molecules and aggregates of molecules are encouraged to be "born" and then to "die."

We do not know to what extent selective pressure can lead these aggregates to become specialized over time. Can some of these molecular aggregates become so elaborate that they can extract energy from their environment and use it to make simple building blocks for proteins and nucleic acids? We will not know the answer to this until we try the experiments.

We will also need new analytical methods to probe the true complexity of the aggregates of molecules that result from these experiments. In particular, we must be able to analyze these messy, complicated collections of molecules and look for the first signs of replication, in the form of the first multimeric molecules that are found more than once in the mixture. Invention of such sensitive analytic techniques will enable us to recreate in the laboratory the appearance of the self-replicating molecules that allowed true Darwinian evolution to begin.

The molecules that we find in our mixtures might be bits of RNA, PNA, other information-carrying molecules with simple backbones, or perhaps even proteinlike molecules. Or they might be something quite different from any of these. We must be prepared for the strong likelihood that more than one type of molecule will be involved in the replication process, just as happens today.

Can all this be done in the timescale of a single scientific career, rather than thousands of years? We think it can, because of the ability

EPILOGUE

of Darwinian selection to speed up the process. It is Darwin who sped up the origin of life from the primordial soup before it was exhausted, and it is Darwin who will aid our efforts to duplicate those events in the laboratory.

These experiments will finally answer the most puzzling question about the origin of life, whether metabolism or genes appeared first. At the 1999 meeting of the International Society for the Study of the Origin of Life, in La Jolla, California, a consensus emerged among the participants that the truth, as it so often does, must lie somewhere in between. But exactly where the truth lies, and exactly how genes and metabolism managed the trick of evolving together, remains to be discovered. We have presented some ideas in this book that we think are testable. You, the reader, are encouraged to think of others.

Top-down approaches are likely to be productive as well. The story that we told in Chapter 9, which shows the pervasiveness and importance of symbiosis in evolution, gives us a clue to how both genes and metabolism could have played a role at the beginning of life. The tree of life that can be constructed from the genealogy of genes tends to be strangely tangled and confusing at its base. The reason for this seems to be that we are seeing the effects of an early period of wild evolutionary experimentation. During this period, evolution was accelerated by symbioses of many different kinds, in which very different protoorganisms came together in various combinations. The result of all this was one kind of genetic material and one overall set of metabolic pathways.

The true nature of the early tree of life is only just beginning to be perceived, thanks to the enormous power of DNA sequencing and computer analysis. Can we see even further back by using the clues buried in our genes? The answer is certainly yes, and the results are sure to be both surprising and illuminating.

In sum, all the chemical and genetic work that we have recounted in this book has shown quite clearly that life on Earth is not some unlikely accident. Life may even have appeared more than once during the early history of our planet. And it is hard to imagine that it has not appeared in many other parts of the universe. Soon, we are confident, it will appear in a test tube. And our world will never be the same.

GLOSSARY

adenosine triphosphate (ATP)   The commonest high-energy molecule in the living cell. This molecule powers many different reactions, ranging from the synthesis of amino acids, proteins, sugars, and fats to the synthesis of the nucleic acids DNA and RNA.

ambiguity reduction   The process by which the ambiguous codon assignments of the early genetic code, in which one codon could specify more than one amino acid, were replaced by more precise assignments. Currently, each codon specifies only one amino acid, although an amino acid can be specified by more than one codon.

amino group   a nitrogen-containing part of an organic molecule, technically known as a sidegroup, which has the formula NH_2.

anaerobe   an organism, such as many bacteria, that can live in the absence of free oxygen. Anaerobes may be *facultative*, indifferent to the presence or absence of oxygen, or *obligate*. Obligate anaerobes are actually poisoned by oxygen. In some cases their ancestors may never have lived in an environment with oxygen, and such obligate anaerobes may be descendants of very primitive organisms.

anaerobic   an environment without free oxygen.

archaea (pl.), archaeon (sing.)   one of the three current major domains of life, the others being the eukaryotes and the eubacteria. These bacterialike organisms often, but not always, are found in extreme environments.

archaebacteria   an old name for a member of the archaea.

asteroid belt   a region between the orbits of Mars and Jupiter in which there are many small bodies left over from the Solar System's formation.

base pairs   complementary pairs of bases, one on each strand of DNA. The pairing rules between bases are strict: A always pairs with T and C always pairs with G.

biosphere   the part of the Earth on which life is possible. The biosphere is now known to extend thousands of feet down into deep rock formations.

blue-green algae   these bacterialike organisms are the inventors of the kind of photosynthesis that produces free oxygen. This invention seems to have taken place only once, for the blue-green algae are also the ancestors of the chloroplasts that are found in higher plants and that are responsible for the photosynthesis on which we all depend for life.

bolide   an asteroid or comet that crashes onto the Earth or another body in the Solar System, generating a huge fire-ball. Comet Shoemaker-Levy broke up into a series of bolides that collided with Jupiter, producing a succession of spectacular explosions.

brown dwarf an object substantially larger than Jupiter but with a mass no more than 40 percent that of the Sun. These objects are not big enough for gravitational collapse to heat them to the point that nuclear reactions can be triggered. It is uncertain exactly what the cutoff is that separates a brown dwarf from a very large planet like Jupiter. It has been speculated that brown dwarfs may be very common in the universe. They may have moons.

carbonaceous chondrite see chondrite.

carboxyl group the acidic component (sidegroup) found on all organic acids and represented by the formula COOH.

chondrite a type of stony meteorite that contains numerous small spherules of silicate (silica, silicon dioxide) minerals. A subset of this type of meteorites, the carbonaceous chondrites, contain several per cent organic carbon.

chromatography a chemical technique in which organic compounds can be separated using columns or sheets of filtering material (see **geochromatography**).

coacervates structures, the existence of which was originally suggested by Aleksandr Oparin, that may have been important in the origin of life. As Oparin envisioned them, they were globules surrounded by membranes that enclosed and protected a variety of organic molecules.

codon a series of three nucleobases along the strand of a DNA or messenger RNA molecule that codes for a particular amino acid in a peptide sequence.

continental drift the slow movement of continental and oceanic plates, measured in a few centimeters per year, that has resulted in massive alterations of the features of the Earth's crust over geologic time. Colliding plates can generate mountain ranges, and upwellings of the mantle can push plates apart to produce ocean basins.

continuously habitable zone (CHZ) the zone around a star in which the conditions are favorable for the existence of life throughout most of the star's lifetime.

coronene a polycyclic aromatic hydrocarbon that contains seven aromatic rings and has the formula $C_{24}H_{12}$.

crust the outermost layer of the Earth. The crust on the continents is thicker (between 30–50 km) than that of the ocean basins (thickness 5–10 km).

cryptomonads small single-celled protozoa, mostly marine, that often have red or green algae living inside them.

dissymmetric a word used by Louis Pasteur to describe the property of an object wherein its mirror images are not superimposable on each other. Examples include left and right hands and L- and D-amino acids. Also see **optical activity**.

Drake equation a formula for working out the number of communicating civilizations in the galaxy.

eubacteria "true" bacteria, as distinguished from archaea. Like archaea they are single-celled organisms that possess chromosomes but lack a cell nucleus. They can be distinguished from archaea because many of their genes are very different from their archaeal equivalents.

eukaryote an organism made up of cells that possess a true cell nucleus. Eukaryotic organisms such as ourselves undergo regular sexual recombination, unlike eubacteria or archaea in which sex is irregular and often does not involve the entire complement of genes.

Gaia hypothesis the idea, first proposed by James Lovelock, that the organisms in the entire biosphere of Earth interact with each other and their physical environment to stabilize the Earth's environmental conditions over a narrow range.

GLOSSARY

gegenschein a faint light, visible on the darkest night, caused by reflection of sunlight from the cloud of small particles in orbit in the inner Solar System. If you lived on Saturn you would see no gegenschein, for the outer Solar System has been swept clear of these tiny particles.

gemisch a messy, complex mixture consisting of many different types of components (a useful word from the Yiddish).

geochromatography a process by which finely divided minerals in the Earth itself have separated out organic compounds, just as chemists use chromatography to do the same thing in the laboratory.

half life the period of time needed for half of a radioactive isotope in a sample to decay.

hydrothermal vent a deep ocean vent at which magma-heated and mineral-laden water spews out at a high rate. The water, heated to temperatures of more than 300°C, remains liquid because of the high pressure.

hydroxyl ion the negatively charged fragment of the water molecule, which has the formula OH⁻.

interferometry an astronomical technique in which the images from two or more telescopes are superimposed. Interferometry has many uses, and one of the most intriguing is its ability to cancel out the light from a point source such as a star by superimposing the images in such a way that the oscillations in light intensity from one image are the reverse of the oscillations from the other image. With this method, planets orbiting the star can be directly observed, and the composition of their atmospheres can be investigated using spectral analyses.

isotopes different forms of an element that have the same number of protons in their nuclei, and thus the same atomic number, but that have different numbers of neutrons and thus different atomic masses. There are two kinds of isotopes, stable and unstable. Isotopes that are unstable are called radioactive and disintegrate at a constant decay rate. Examples of stable isotopes include carbon-12 and carbon-13. Carbon-14, and uranium-238 and -235, are examples of unstable isotopes.

light-year the distance, 9.4607 trillion kilometers or 5.8786 trillion miles, that light travels in one Earth year.

mantle the layer of the Earth that lies between the crust and the core. The mantle is so hot that the rock that makes it up is molten, but pressure is so great that it flows only slowly. Convection currents in the mantle are the cause of continental drift.

messenger RNA these single-stranded nucleotide chains carry the information from the genes to the part of the cell outside the nucleus, where this information is used to manufacture proteins.

microfossils fossils of unicellular organisms that can only be seen under the microscope in specially prepared rock samples.

micrometeorite a particle from space that is small enough to be slowed down when it reaches the Earth's atmosphere without being burnt up. Approximately 50 micrometeorites per square meter fall to Earth each day.

microtubules long tubular chains of protein molecules that give much of the structure to eukaryotic cells.

mitochondria small structures in most eukaryotic cells, the descendants of free-living bacteria, that are chiefly responsible for the manufacture of ATP and other energy-rich compounds.

multimer a polymer of moderate length made of several repeating similar units, such as a short peptide or nucleic acid sequence.

GLOSSARY

mycoplasmas   the smallest independently living cells at the present time. Mycoplasmas, small as they are, have over 500 genes.

nebular hypothesis   a general theory that describes how stars and their associated solar systems are formed from the condensation of clouds of dust and gas in space.

Oort cloud   a vast collection of bodies made up of rock and ice that orbit the Sun at a distance starting in the region beyond the orbit of Pluto and extending out to nearly 1.5 light-years. Unlike the bodies of the Solar System, which orbit the Sun in roughly the same plane, the Oort cloud objects form a vast sphere around the Sun. It is estimated that billions of objects exist in this region, and there is evidence that this is the place where most comets originate.

optical activity   the property of some crystals, gases, liquids, and solutions to rotate plane-polarized light to the left or right. It occurs because the molecules that make up the substance through which the light beam is shone are asymmetric, i.e., they have no plane of symmetry. Asymmetric molecules are mirror images of each other that cannot be superimposed. This asymmetric property is also referred to as handedness. Examples are the L- and D-forms of amino acids.

panspermia   the theory that life can arrive on a planet from a spore, seed, or cell derived from life elsewhere in the universe.

peptide nucleic acid (PNA)   an artificial molecule constructed to look as much like a nucleic acid as possible.

phosphorylation   the attachment of a phosphate group (H_2PO_4) to another organic molecule. The phosphorylation of ADP to ATP is difficult to do, and adds a great deal of stored energy to the molecule.

planetesimal   small objects of condensed rocky material that orbited the Sun when it was young, and aggregated to form the terrestrial planets.

plasmids   small, usually circular, DNA molecules that can carry a number of genes from one species of bacterium to another. Other kinds of such transposable elements can pass genes back and forth among eukaryotic species.

plate tectonics   the theory that the Earth's continental and oceanic crust and outermost portion of the mantle is fractured into large plates that move relative to each other. Convective currents in the mantle provide the driving force for this motion. The plate motion is responsible for global mountain building, earthquake activity, and volcanism, all of which are most pronounced along plate boundaries (see **continental drift**).

polycyclic aromatic hydrocarbons (PAHs)   molecules that are made up only of carbon and hydrogen and that have two or more benzenelike structures. These molecules, some of which are enormously complex, are found in oil deposits at the present time and are also found in carbonaceous chondrites.

polymerase chain reaction (PCR), polymerase chain-reaction technique   a method for making many copies of a gene or other segment of DNA.

prokaryotes   organisms that have no cell nucleus but do have chromosomes. They include the bacteria, blue-green algae, and archaea.

protobionts   the first complex structures that arose during the formation of life that were capable of reproduction. Although protobionts can be considered the first living cells, they were very different from present-day cells and probably only parts of them were capable of accurate replication.

proton   a positively charged hydrogen nucleus, often derived from a water molecule, with the formula H^+.

GLOSSARY

pulsar a rapidly rotating object, strongly suspected to be a neutron star (an extremely dense object made up mainly of neutrons and a few kilometers in diameter). Pulsars emit beams of radio waves or other types of electromagnetic radiation, which we perceive as pulses as these objects whirl, sometimes at hundreds of revolutions a second.

racemic mixture a mixture that contains exactly equal amounts of the asymmetric forms of an optically active molecule (see **optical activity**). Such a mixture does not cause plane-polarized light to rotate in either direction.

radiodating the technique of dating rocks and minerals using the ratios of various radioactive and nonradioactive minerals.

reticulate evolution a process in which genes are passed from one species to another. Sometimes the recipient of these genes is very distantly related to the donor. When family trees are built using these genes, the results can often be very confusing.

ribozyme a molecule of RNA that has catalytic activity. Ribozymes are rare in living cells today, but may have been much more common when life began.

ribosomes structures in the cell that translate information from messenger RNA into proteins. Ribosomes are the factories that make new parts of cells as they grow.

spirochete a small spiral-shaped bacterium. Spirochetes are the cause of human disease such as syphilis and yaws, and their harmless relatives are common denizens of the soil. It was thought that ancestors of spirochetes might have contributed to eukaryotic cells through symbiosis, but this is almost certainly not correct.

supernova an explosion of a massive star (mass greater than 8 to 10 times the mass of the Sun) which ejects, at speeds about a tenth the speed of light, most of the original mass into space. The shell of material left behind may form a pulsar or black hole, a region where gravity is so strong that no radiation or matter can escape.

symbiosis "living together," two different organisms cooperating to make a more effective competitor than either one by itself.

thioester the combination of a carboxyl group of an organic acid with a thiol (SH) to yield a compound part of which has the chemical formula COSH.

ventists researchers who speculate that the synthesis of prebiotic compounds that were required for the origin of life, or even the origin of life itself, took place within deep ocean hydrothermal vents or in their vicinity.

zircon a silicate mineral, diamondlike in appearance, that contains silica silicon dioxide, and the element zirconium. Some zircons in meteorites contain material from the period before the Solar System formed.

NOTES AND REFERENCES

Introduction

p. xii. Pedersen's account of the bacterial life he has found in Swedish rocks is in K. Pedersen, "Microbial Life in Deep Granitic Rock," *FEMS Microbiology Reviews* 20 (1997): 399–414.

p. xii. Creation myths of many different religions have been gathered together in an excellent web site, the Encyclopedia Mythica (http://www.pantheon.org/mythica/).

p. xiii. The tragic story of Giordano Bruno is told in Hilary Gatti, *Giordano Bruno and Renaissance Science* (Ithaca: Cornell University Press, 1999).

p. xv. Dyson's book is Freeman J. Dyson, *Origins of Life* (New York: Cambridge University Press, 1985).

p. xvi. Fascinating accounts of the tenacity of life in extreme environments can be found in Michael Gross, *Life on the Edge: Amazing Creatures Thriving in Extreme Environments* (New York: Plenum, 1998). The story of seals taking back the beaches of La Jolla can be found in Donna Foote, "Yikes, We're Sealed In!" in *Newsweek,* December 14, 1998.

Chapter 1: The Rise and Fall of Spontaneous Generation

p. 1. An excellent source on Redi's life and experiments is the introduction and text of Francesco Redi, *Experiments on the Generation of Insects* (Chicago: Open Court, 1909).

p. 1. The history and politics behind the spontaneous-generation debate is recounted in John Farley, *The Spontaneous Generation Controversy from Descartes to Oparin* (Baltimore: Johns Hopkins University Press, 1977).

p. 2. Aristotle's views on the origin of living things from the four basic elements are found in his Generation and Corruption, Book 2. The quote from van Helmont is recounted in Willy Ley, *Exotic Zoology* (New York: Viking, 1959), p. 241. Kircher's claims are recounted in Joscelyn Godwin, *Athanasius Kircher: A Renaissance Man and the Quest for Lost Knowledge* (London: Thames and Hudson, 1979).

p. 6. An accessible account of van Leeuwenhoek's discoveries is found in Brian J. Ford, *Single Lens: The Story of the Simple Microscope* (London: Heinemann, 1985).

p. 11. John C. Greene, *The Death of Adam: Evolution and Its Impact on Western Thought* (Ames, Iowa: Iowa State University Press, 1959), gives a detailed account of Lamarck's ideas.

p. 12. Virchow's remarkable life is examined in Byron A. Boyd, *Rudolf Virchow: The Scientist as Citizen* (New York: Garland, 1991).

pp. 15–19. Excellent sources on both Biot's and Pasteur's research of optical activity can be found in J. Applequist, "Optical Activity: Biot's Bequest," *American Scientist* 75 (1987): 59–68, and G. L. Geison and J. A. Secord, "Pasteur and the Process of Discovery: The Case of Optical Isomerism," *ISIS* 79 (1988): 6–36. The exact English translation of Biot's exclamation upon seeing the different rotation of the left- and right-handed tartaric acid solutions varies considerably. We have used the one found in Geison and Secord. For readers interested in trying to repeat Pasteur's experiment with tartaric acid crystals, see G. B. Kaufman and R. D. Myers, "The Resolution of Racemic Acid: A Classic Stereochemical Experiment for the Undergraduate Laboratory," *Journal of Chemical Education* 12 (1975): 777–781.

p. 23. Quote from *Oeuvres de Pasteur*, edited by Pasteur Vallery-Radot (Paris: Masson et Cie, 1922), 2:332–333.

p. 26. Because radioactivity was not discovered until after the turn of the century, Thomson was unaware of this internal source of the Earth's heat. For an account of the age-of-the-Earth debate during the later part of the nineteenth century, see D. Fischer, "The Time They Postponed Doomsday," *New Scientist* June 6, 1985, pp. 39–43.

pp. 28-29. Thomson suggested in the lecture that during the collision of planets and other bodies in the cosmos, debris must have blasted off their surfaces and ended up in space. This insightful idea is now widely accepted and is used to explain how pieces of Mars have ended up on Earth. The poem "No Conjuror's Conjection" was brought to our attention in an article by C. Pillinger and J. Pillinger, "A Brief History of Exobiology or There's Nothing New in Science," *Meteoritics and Planetary Science* 32 (1997): 443–446.

p. 31. We now know that this class of fairly rare carbon-containing meteorites, called carbonaceous chondrites, contains a rich assortment of organic compounds, some of which play a major role in biochemistry.

p. 31. Arrhenius was the first to show the relationship between the rate of a chemical reaction and temperature. His famous "Arrhenius" equation indicates that the lower the temperature, the slower the speed of the reaction. Thus, he calculated that below −200°C, cells would retain their "vital functions" for a period longer than that required to reach other worlds.

Chapter 2: Primordial Soup

p. 35. This book titled *Proiskhozhdenie zhizny* was not widely available in English until 1967, when it was published as part of a book by J. D. Bernal. This article, as well as the ones by Haldane, Miller, Urey, and others discussed in this part and other parts of the book can by found in D. W. Deamer and Gail R. Fleischaker, *Origins of Life: The Central Concepts* (Boston: Jones and Bartlett Publishers, 1994).

p. 36. Haldane was not aware of Oparin's 1924 book. At a meeting on the origin of life in 1963, he generously gave Oparin the credit: "I have little doubt that Professor

Oparin has the priority over me." He continued with a bite of self-deprecation, noting that if he had read the book, the question would perhaps not be one of priority, but "The question of plagiarism might." A discussion of the priority issue can be found in *The Origins of Prebiological Systems*, edited by S. W. Fox (New York: Academic Press, 1965), pp. 96–98.

p. 37. Why did the origin-of-life hypothesis originate with Oparin and Haldane at this particular time? The two scientists came from what might appear to be totally different backgrounds, but they had a great deal in common. Oparin was a young scientist in a country recently taken over by communists, and Haldane was a member of a distinguished scientific family centered in Cambridge, England. They were both Marxists, Haldane by choice and Oparin because he was required to be one in order to survive in Soviet Russia. It is tempting to speculate that the materialist worldview of Marxism encouraged both Oparin and Haldane to think about subjects that, in earlier religion-dominated cultures, would have been considered dangerously heretical. But although Haldane had no problems, Oparin actually had great difficulty publishing his ideas. His thesis, which incorporated his model, was rejected by his professors. Only later was he embraced by the communist regime.

The politics of both Haldane and Oparin strongly influenced their behaviors, and the result has been considerable controversy. As we look at their careers, we must remember that they were both prisoners of their time.

Even though he was one of the leading geneticists and biological geniuses of this century, Haldane never published any criticism of the anti-Darwin plant breeder Trofim Lysenko, who with Stalin's blessings essentially destroyed all genetics research in the USSR for decades. Geneticists in the West were well aware that Lysenko's theories were nonsense and, even worse, that Lysenko had helped to send dozens of scientists to the Gulag. Haldane's silence about the Lysenko affair has damaged his reputation. On the other hand, Haldane was quite capable of endearing and dramatic gestures for causes he believed in. At fifty-nine, he resigned his professorship at University College, London, and moved to India, becoming an Indian citizen and aiding greatly in the establishment of a science infrastructure there.

Oparin's case is even more politically charged. We cannot imagine the pervasive political atmosphere in which he did his work. When his first book was published in 1924, the publisher added the slogan "Workers of the world, unite!" to the cover, something that was standard practice in those days. In his books, Oparin makes frequent reference to the pseudobiological writings of Friedrich Engels and occasionally to Lenin as well, because this was the politically correct thing to do. Oparin eventually became a member of the Politburo, but he was never a member of the Communist party.

Far more discomfitingly, Oparin was a supporter and an acquaintance of Lysenko, and this helped him to prosper under the Stalinist regime. Immediately after Stalin publicly embraced Lysenko's ideas in 1948, Oparin was appointed to the position of secretary of the Biochemical Section of the USSR Academy of Science, where, according to historian Loren R. Graham (*Science, Philosophy, and Human Behavior in the Soviet Union* [New York: Columbia University Press, 1987]), he "administered the liquidation of genetics." There is no doubt he acquiesced in, and may even have been directly involved in, the exile of a number of scientists to the Gulag.

Oparin's Marxist history has made even the use of his name controversial. The International Society for the Study of the Origin of Life (ISSOL) presents a medal periodically in his honor. Some ISSOL members have recently objected to naming some of their awards after Oparin, because of his apparent involvement in the oppression and persecution associated with Stalin's regime. Others in ISSOL defend the Oparin medal, pointing out that scientific contributions must be separated from personal behavior. This latter argument is akin to that used by the U.S. senators who voted against the impeachment of President Clinton even though they found his behavior repugnant.

Oparin's case, like that of the brilliant physicist Werner Heisenberg, who ran the atom bomb project for Nazi Germany, is a difficult one for another reason. How much of his support for the Stalin regime was the result of fear for his safety and life? It is hard for us, secure in our cocoon of freedom, to imagine the corrosive effects of such fear. In an interview with Graham in 1971, Oparin said, "It is easy for you, an American, to make such accusations. If you had been here in those years, would you have had the courage to speak out and be imprisoned in Siberia?"

True enough. After all, when Galileo was shown the instruments of torture by the Inquisition, he recanted his beliefs. How many of us would not? Still, one wishes that Oparin had not so openly and apparently enthusiastically embraced and participated in Stalin's terrorist tactics.

We cannot set these grim facts completely aside. Nonetheless, Oparin's 1924 and 1936 books laid the foundation for origin-of-life research in this century. The ISSOL Oparin medal is a small recognition of these immense contributions.

p. 40. The description of Miller's lecture and how he ended up doing the experiment is partly based on S. L. Miller, "The First Laboratory Synthesis of Organic Compounds under Primitive Earth Condition," in *The Heritage of Copernicus: Theories "Pleasing to the Mind,"* edited by J. Neyman (Cambridge, Mass.: MIT Press, 1974), pp. 228–242. We obtained additional details during conversations with Miller. It is unclear exactly who asked Miller the question about whether something similar to his experiment really happened on the early Earth (p. 58). According to Urey's National Academy of Sciences biography, the question is attributed to Enrico Fermi. But Miller told us he does not think Fermi asked the question, because he would have recognized Fermi's distinctive accent.

p. 41-42. Urey was evidently unaware of a 1938 report by Groth and Suess, the latter of whom became a close collaborator with Urey. They synthesized formaldehyde by the ultraviolet irradiation (produced by a xenon arc lamp) of carbon dioxide and water (for the Groth and Suess paper, see the Deamer and Fleischaker reference given for p. 52). Even this paper may have been anticipated by one published in 1921 in which similar results were apparently obtained.

p. 43. For the classic encyclopedic paper on the synthesis of the elements, see E. M. Burbidge et al., "Synthesis of Elements in Stars," *Reviews of Modern Physics* 29: (1957) 547–650. Of the four authors, only Fowler received the Nobel prize.

p. 47. Almost all scientific papers are evaluated by two knowledgeable referees who are supposed to judge the validity of the results. It is not uncommon to have reviewers who balk at new and unexpected results. The reviewer later identified himself and apologized to Miller for the delay.

p. 50. The first publications confirming Miller's results were L. Hough and A. F. Rogers, "Synthesis of Amino Acids from Water, Hydrogen, Methane and Ammonia," *Journal of Physiology* 132 (1956): 28P–30P, and P. H. Abelson, "Amino Acids Formed in 'Primitive Atmospheres,'" *Science* 124 (1956): 935.

NOTES AND REFERENCES

p. 54. The Fox quote can be found in W. H.-P. Thiemann, "Another Short Hommage à Sidney W. Fox from a Humble Disciple," *Newsletter of ISSOL (International Society for the Study of the Origin of Life)* 26 (1999): 15–16.

Chapter 3: The Earth's Apocalyptic Beginnings

p. 60. An excellent account of the search for evidence of life in old rocks can be found in J. William Schopf, *Cradle of Life: The Discovery of Earth's Earliest Fossils* (Princeton, N.J.: Princeton University Press, 1999).

p. 61. Abelson's paper, "Chemical Events on the Primitive Earth," was published in *Proceedings of the National Academy of Sciences (U.S.)* 55 (1966): 1365–1372.

p. 64. A discussion of the nebular hypothesis, as well as alternatives, can be found in Stephen G. Brush, *Nebulous Earth: The Origin of the Solar System and the Core of the Earth from Laplace to Jeffreys* (Cambridge: Cambridge University Press, 1996).

p. 67. In a lecture in 1904 before the Royal Institution of Great Britain, Ernest Rutherford made one of the first attempts to use radioactive decay to tackle the question of the Earth's age. Thomson, now Lord Kelvin, was in the audience, but slept through most of the lecture, although he awoke near the end, when the issue of the Earth's age was addressed. Rutherford's age estimate of nearly a billion years must have been unsettling to Kelvin, who was not used to being told he was wrong. Kelvin and Rutherford cordially debated radioactivity and the age of the Earth at Lord Rayleigh's estate after the lecture. Until Kelvin died on December 17, 1907, however, he still held fast to his 24-million-year age of the Earth, refusing to accept that radioactive decay explained the increasing temperature with depth in the Earth.

p. 69. For another summary of the efforts to determine the age of the Earth, besides the book by Dalrymple, see S. G. Brush, *Transmuted Past: The Age of the Earth and the Evolution of the Elements from Lyell to Patterson* (Cambridge: Cambridge University Press, 1996).

p. 70. For the most recent summary of the age of the Moon, see D.-C. Lee et al., "Age and Origin of the Moon," *Science* 278 (1997): 1098–1103. For the first paper on the formation of the Moon by the collision of a giant impactor with the young Earth, see W. K. Hartmann and D. R. Davis, "Satellite-Sized Planetesimals and Lunar Origin," *Icarus* 24 (1975): 504–515.

p. 76. A discussion of the primitive, smoggy atmosphere can be found in C. Sagan and C. Chyba, "The Early Faint Sun Paradox: Organic Shielding of Ultraviolet-Labile Greenhouse Gases," *Science* 276 (1997): 1217–1221.

p. 77. Currently, rivers discharge about a billion liters of water into the oceans per second. The total volume of the Earth's oceans is 10^{21} liters.

p. 78. The impact frustration hypothesis can be found in K. A. Maher and D. J. Stevenson, "Impact Frustration of the Origin of Life," *Nature* 331(1988): 612–614. Also see N. H. Sleep et al., "Annihilation of Ecosystems by Large Asteroid Impacts on the Early Earth," *Nature* 342 (1989): 139–142.

p. 80. The analogy of atmospheric gases with how a greenhouse traps solar energy was first recognized by the French scientist Jean-Baptiste Fourier in 1827. We now know the situation is much more complex than this simple analogy implies. Air in a greenhouse cannot mix with the outside air, and this lack of circulation plays an important role in keeping the interior warm.

p. 83. James Lovelock put forward his Gaia hypothesis at length in *Gaia: A New Look at Life on Earth* (Oxford: Oxford University Press, 1979). The hypothesis has been harshly criticized since, but a consensus is slowly emerging that life does have an influence on moderating and perhaps stabilizing conditions on the Earth. For thoughts about how natural selection could play a role in such stabilization, see Timothy M. Lenton, "Gaia and Natural Selection," *Nature* 394 (1998): 439–447.

Chapter 4: Prebiotic Soup: The Recipe

p. 85. For an account of the organized element controversy, see B. Nagy, *Carbonaceous Meteorites* (Amsterdam, N.Y.: Elsevier Scientific Publishing, 1975), and F. W. Fitch, H. P. Schwarcz, and E. Anders, "'Organized Elements' in Carbonaceous Chondrites," *Nature* 193 (1962): 1123–1126.

p. 86. For the first report on amino acids in the Murchison meteorite, see K. Kvenvolden et al., "Evidence for Extraterrestrial Amino Acids and Hydrocarbons in the Murchison Meteorite," *Nature* 228 (1970): 923–926.

p. 89. For a discussion of the organics-from-space concept, see C. Chyba and C. Sagan, "Endogenous Production, Exogenous Delivery and Impact-Shock Synthesis of Organic Molecules: An Inventory for the Origin of Life," *Nature* 355 (1992): 125–132. A debate on the importance of organic material from space can be found in S. L. Miller and C. Chyba, "Whence Came Life?" *Sky and Telescope* June (1992): 604–605.

p. 95. For the oil-slick concept, see A. C. Lasaga, H. D. Holland, and M. J. Dwyer, "Primordial Oil Slick," *Science* 174 (1971): 53–55.

p. 96. For a discussion of the theory that life arose in hydrothermal vents, see J. B. Corliss, J. A. Baross, and S. B. Hoffman, "An Hypothesis Concerning the Relationship Between Submarine Hot Springs and the Origin of Life on Earth," *Oceanologica Acta* (1981): 59–69 and see S. Simpson, "Life's First Scalding Steps," *Science News* 155 (1999): 24–26.

p. 101. On the mineral origins of life, see A. G. Cairns-Smith, *Genetic Takeover and the Mineral Origins of Life* (Cambridge: Cambridge University Press, 1982). Also see A. G. Cairns-Smith, *The Life Puzzle: On Crystals and Organisms and on the Possibility of Crystal as an Ancestor* (Edinburgh, Scotland: Oliver and Boyd, 1971).

Chapter 5: Sorting Out the Gemisch

p. 110. Bergson's philosophy of evolutionary change is set out in Henri Bergson, *Creative Evolution* (Westport, Conn.: Greenwood Press, 1975 [1911]).

p. 117. A discussion of geochromatography and its role in the natural separation of organic compounds can be found in M. R. Wing and J. L. Bada, "Geochromatography on the Parent Body of the Carbonaceous Chondrite Ivuna," *Geochimica et Cosmochimica Acta* 55 (1991): 2937–2942.

Chapter 6: The First Protobionts

p. 123. Also see J. D. Bernal, *The Origin of Life* (Cleveland: World Publishing, 1967). An account of Bernal's life is found in M. Goldsmith, *Sage: A Life of J. D. Bernal* (London: Hutchinson, 1980).

p. 125. The work of Cech and Altman is explained in T. R. Cech, "Self-Splicing and Enzymatic Activity of an Intervening Sequence RNA from *Tetrahymena*," *Bioscience Reports* 10 (1990): 239–260, and C. Guerrier-Takada et al., "The RNA Moiety of Ribonuclease P Is the Catalytic Subunit of the Enzyme," *Cell* 35 (1983): 849–857.

pp. 126-127. An up-to-date summary of the implications of an RNA world in found in R. F. Gesteland, T. R. Cech, and J. F. Atkins, eds., *The RNA World: The Nature of Modern RNA Suggests a Prebiotic RNA* (Cold Spring Harbor, N.Y.: Cold Spring Harbor Press, 1999). Spiegelman's early evolution experiment is described in D. R. Mills, R. L. Peterson, and S. Spiegelman, "An Extracellular Darwinian Experiment with a Self-Duplicating Nucleic Acid Molecule," *Proceedings of the National Academy of Sciences(U.S.)* 58 (1967): 217–224.

p. 128. Eigen's experiments on the Qß replicase system are presented in C. K. Biebricher, M. Eigen, and J. S. McCaskill, "Template-Directed and Template-Free RNA Synthesis by Q-Beta Replicase," *Journal of Molecular Biology* 231 (1993): 175–179.

p. 130. Orgel's prediction can be found in L. E. Orgel, "The Origin of Life on the Earth," *Scientific American* 271 (1994): 77–83.

p. 130. The discovery of Neanderthal DNA is detailed in M. Krings et al., "Neanderthal DNA Sequences and the Origin of Modern Humans," *Cell* 90 (1997): 19–30. Discovery of the 1918 influenza virus gene can be found in A. H. Reid et al., "Origin and Evolution of the 1918 'Spanish' Influenza Virus Hemagglutinin Gene," *Proceedings of the National Academy of Sciences (U.S.)* 96 (1999): 1651–1656. Lindahl's criticism of the RNA world is in T. Lindahl, "Instability and Decay of the Primary Structure of DNA," *Nature* 362 (1993); also see Lindahl's paper, "Facts and Artifacts of Ancient DNA," in *Cell* 90 (1997): 1–3.

p. 131. Joyce's work on catalytic DNA and his suggestion about deconstructing the RNA molecule can be found in S. W. Santoro and G. F. Joyce, "A General Purpose RNA-Cleaving DNA Enzyme," *Proceedings of the National Academy of Sciences (U.S.)* 94 (1997): 4262–4266.

p. 132. The invention of PNAs is chronicled in P. E. Nielsen et al., "Sequence-Selective Recognition of DNA by Strand Displacement with a Thymine-Substituted Polyamide," *Science* 254 (1991): 1497–1500.

p. 133. PNA as a primordial genetic material was explored in P. E. Nielsen, "Peptide Nucleic Acid (PNA): A Model Structure for the Primordial Genetic Material?" *Origins of Life and Evolution of the Biosphere* 23 (1993): 323–327.

p. 134. Miller's conclusion that PNAs are very plausible early genes can be found in S. L. Miller, "Peptide Nucleic Acids and Prebiotic Chemistry," *Nature Structural Biology* 4 (1997): 167–169.

p. 136. Autocatalytic peptides were reported by D. H. Lee, et al., "A Self-Replicating Peptide," *Nature* 382 (1996): 525–528. Other possible catalytic molecules are discussed in E. K. Wilson, "Go Forth and Multiply," *Chemical and Engineering News*, December 7, 1998, pp. 40–44.

p. 137. Eigen suggested the principle of hypercycles in M. Eigen and P. Schuster, "The Hypercycle: A Principle of Natural Self-Organization. Part A: Emergence of the Hypercycle," *Naturwissenschaften* 64 (1977): 541–565. The idea was strongly criticized in U. Niesert, D. Harnasch, and C. Bresch, "Origin of Life Between Scylla and Charybdis," *Journal of Molecular Evolution* 17 (1981): 348–353.

NOTES AND REFERENCES

p. 138. Lancet's computer models are explored in D. Segre et al., "Graded Autocatalysis Replication Domain (GARD): Kinetic Analysis of Self-Replication in Mutually Catalytic Sets," *Origins of Life and Evolution of the Biosphere* 28 (1998): 501–514. Orgel's objections to Dyson's ideas are set out in L. E. Orgel, "The Origin of Life: A Review of Facts and Speculations," *Trends in Biochemical Sciences* 23 (1998): 491–495.

Chapter 7: From Top to Toe

p. 143. The unusual nature of our genetic code is explored in S. J. Freeland and L. D. Hurst, "The Genetic Code Is One in a Million," *Journal of Molecular Evolution* 47 (1998): 238–248.

p. 145. Fitch's approach to the evolution of the code is set out in detail in W. M. Fitch and K. Upper, "The Phylogeny of tRNA Sequences Provides Evidence for Ambiguity Reduction in the Origin of the Genetic Code," *Cold Spring Harbor Symposium in Quantitative Biology* 52 (1987): 759–767.

p. 149. A very thorough survey of the remarkable intracellular structures called mitochondria can be found in I. E. Scheffler, *Mitochondria* (New York: Wiley-Liss, 1999).

p. 150. Some of the story of Peter Mitchell's remarkable career is recounted in P. Aldhous, "One Man's Achievement," *Nature* 347 (1990): 605, and P. C. Hinkle and K. D. Garlid, "Peter Mitchell: 1920–1992," *Trends in Biochemical Sciences* 17 (1992): 304–305.

p. 154. Some background on the discovery of the structure of ATPase is laid out in C. Surridge, "Nobel Prizes Honour Biologists' Work on Protein Energy Converters," *Nature* 389 (1997): 771.

p. 155. David Deamer has summarized his work on coacervates and early energy-generating systems in D. W. Deamer, "The First Living Systems: A Bioenergetic Perspective," *Microbiology and Molecular Biology Reviews* 61 (1997): 239–261.

Chapter 8: Journey to the Center of the Earth

p. 160. A brief description of the microbial White Sands community is found in J. James et al., "Isolation and Characterization of Cyanobacterial Strains from the Gypsum Dunes of the White Sands National Monument," *Abstracts of the General Meeting of the American Society for Microbiology* 94 (1994): 354.

p. 162. The more complex microbial community of the Camargue is explored in P. Caumette et al., "Microbial Mats in the Hypersaline Ponds of Mediterranean Salterns (Salins-de-Giraud, France)," *FEMS (Federation of European Microbiological Societies) Microbiology Ecology* 13 (1994): 273–286.

p. 166. A discussion of the microbiology of desert varnish can be found in M. Eppard et al., "Morphological, Physiological, and Molecular Characterization of Actinomycetes Isolated from Dry Soil, Rocks, and Monument Surfaces," *Archives of Microbiology* 166 (1996): 12–22. The slow development of varnish in the Australian desert is chronicled in D. Dragovich, "Fire, Climate, and the Persistence of Desert Varnish near Dampier, Western Australia," *Palaeogeography Palaeoclimatology Palaeoecology* 111 (1994): 279–288.

p. 168. The destruction and recolonization of a vent community is chronicled in V. Tunnicliffe et al., "Biological Colonization of New Hydrothermal Vents Following an

Eruption on Juan de Fuca Ridge," *Deep-Sea Research: Part I, Oceanographic Research Papers* 44 (1997): 1627–1643. Clam–bacterial symbiosis off the coast of California is examined in J. J. Childress et al., "The Role of a Zinc-Based, Serum-Borne Sulfide-Binding Component in the Uptake and Transport of Dissolved Sulfide by the Chemoautotrophic Symbiont-Containing Clam *Calyptogena elongata*," *Journal of Experimental Biology* 179 (1993): 131–158.

p. 169. Gold's calculation of subterranean bacterial mass is found in T. Gold, "The Deep Hot Biosphere," *Proceedings of the National Academy of Sciences (U.S.)* 89 (1992): 6045–6049. Lipman's early report on bacteria in coal is in C. B. Lipman, "Living Microorganisms in Ancient Rocks," *Journal of Bacteriology* 22 (1931): 183–198.

p. 170. Gold recounts his discovery in a book with the same title as his earlier paper: T. Gold, *The Deep Hot Biosphere* (New York: Copernicus, 1999). For a critique of this book and Gold's disocvering oil, see K. Peters, Organic Geochemistry, vol. 30, pp. 473-475, 1999.

p. 171. Discoveries of very deep bacterial communities are summarized in K. Pedersen, "The Deep Subterranean Biosphere," *Earth-Science Reviews* 34 (1993): 243–260.

p. 172. T. O. Stevens and J. P. McKinley, "Lithoautotrophic Microbial Ecosystems in Deep Basaltic Aquifers," *Science* 270 (1995): 450–454, is the primary reference for the work on the deep microbial communities near Hanford.

p. 174. Discovery of this unusually self-sufficient bacterium is recounted in J. G. Zeikus and R. S. Wolfe, "*Methanobacterium thermoautotrophicus* sp. n., an Anaerobic, Autotrophic, Extreme Thermophile," *Journal of Bacteriology* 109 (1972): 707–715.

p. 175. Pace's new version of the tree of life is given in N. R. Pace, "A Molecular View of Microbial Diversity and the Biosphere," *Science* 276 (1997): 734–740.

p. 177. Paul Rabinow, in *Making PCR: A Story of Biotechnology* (Chicago: University of Chicago Press, 1996), gives some of the scientific and political story behind the seminal invention of this technique.

p. 179. The story of the discovery of the archaea, and the price paid by their discoverer, is found in Virginia Morell, "Microbiology's Scarred Revolutionary," *Science* 276 (1997): 699–702.

p. 180. T. R. Holoman et al., "Characterization of a Defined 2,3,5,6-Tetrachloro-biphenyl-ortho-dechlorinating Microbial Community by Comparative Sequence Analysis of Genes Coding for 16S rRNA," *Applied and Environmental Microbiology* 64 (1998): 3359–3367, gives details of the discovery of archaea in estuarine mud.

Chapter 9: Evolution by Committee

p. 183. Lynn Margulis recounts her early career in Lynn Margulis, *Symbiotic Planet: A New Look at Evolution* (New York: Basic Books, 1998). She traces the history of the symbiosis idea in Lynn Margulis, *Origin of Eukaryotic Cells: Evidence and Research Implications for a Theory of the Origin and Evolution of Microbial, Plant, and Animal Cells on the Precambrian Earth* (New Haven: Yale University Press, 1970).

p. 185. Alternative trees of early life that can be traced using various molecules are discussed in W. F. Doolittle and J. M. Logsdon, Jr., "Archaeal Genomics: Do Archaea Have a Mixed Heritage?" *Current Biology* 8 (1998): R209–211. Woese's view about genetic anarchy at the beginning of life is in C. Woese, "The Universal Ancestor," *Proceedings of the National Academy of Sciences (U.S.)* 95 (1998): 6854–6859.

p. 189. Carroll's studies on the evolution of insect flight are found in S. B. Carroll, S. D. Weatherbee, and J. A. Langeland, "Homeotic Genes and the Regulation and Evolution of Insect Wing Number," *Nature* 375 (1995): 58–61. The oldest winged insect fossils were examined by J. Kukalova-Peck and C. Brauckmann, "Wing Folding in Pterygote Insects, and the Oldest Diaphanopterodea from the Early Late Carboniferous of West Germany," *Canadian Journal of Zoology* 68 (1990): 1104–1111. Some of the latest information on the sequence of *E. coli* O157:H7 is found in G. Plunkett III et al., "Sequence of Shiga Toxin 2 Phage 933W from *Escherichia coli* O157:H7: Shiga Toxin as a Phage Late-Gene Product," *Journal of Bacteriology* 181 (1999): 1767–1778. Fred Blattner's group is currently sequencing the entire genome of this dangerous pathogen.

p. 191. L. Margulis and R. Fester, eds., *Symbiosis as a Source of Evolutionary Innovation: Speciation and Morphogenesis* (Cambridge, Mass.: MIT Press, 1991), contains summaries of many amazing symbioses found throughout every branch of the tree of life.

p. 193. Recent work on Jeon's remarkable bacterial–amoeba system is found in J. W. Pak and K. W. Jeon, "A Symbiont-Produced Protein and Bacterial Symbiosis in *Amoeba proteus*," *Journal of Eukaryotic Microbiology* 44 (1997): 614–619.

p. 196. S. E. Douglas et al., "Cryptomonad Algae Are Evolutionary Chimaeras of Two Phylogenetically Distinct Unicellular Eukaryotes," *Nature* 350 (1991): 148–151, gives details on the nature of the cryptomonad symbiosis.

p. 198. Doolittle's paper is R. F. Doolittle et al., "Determining Divergence Times of the Major Kingdoms of Living Organisms with a Protein Clock," *Science* 271 (1996): 470–477, and the immediate critical response to it is summarized in V. Morell, "Proteins 'Clock' the Origins of All Creatures—Great and Small," *Science* 271 (1996): 448.

p. 203. Iwabe's important paper documenting the early history of duplicated genes is N. Iwabe et al., "Evolutionary Relationship of Archaebacteria, Eubacteria, and Eukaryotes Inferred from Phylogenetic Trees of Duplicated Genes," *Proceedings of the National Academy of Sciences (U.S.)* 86 (1989): 9355–9. Li's dating of the duplication events is in R. M. Adkins and W-H. Li, "Dating the Age of the Last Common Ancestor of All Living Organisms with a Protein Clock," in *Horizontal Gene Transfer*, edited by M. Syvanen and C. I. Kado (New York: Chapman and Hall, 1998).

p. 205. Doolittle's reexamination of the problem is found in D. F. Feng, G. Cho, and R. F. Doolittle, "Determining Divergence Times with a Protein Clock: Update and Reevaluation," *Proceedings of the National Academy of Sciences (U.S.)* 94 (1997): 13028–13033.

Chapter 10: Life Elsewhere

p. 211. The quote by Huygens can be found in C. Sagan, *Cosmos* (New York: Random House, 1980), p. 162.

p. 215. A good review of SETI and the Drake equation can be found in S. J. Dick, *Life on Other Worlds* (Cambridge: Cambridge University Press, 1998), pp. 200–235.

p. 219. A discussion of the Moon hoax can be found in W. Ley, *Watchers of the Skies: An Informal History of Astronomy from Babylon to the Space Age* (London: Sidgwick and Jackson, 1963), pp. 268–275.

p. 221. For a description of the Venusian organisms, see H. Morowitz and C. Sagan, "Life in the Clouds of Venus?" *Nature* 215 (1967): 1259–1260. Sagan depicted similar organisms living in the atmosphere of a Jupiter-like planet in his 1980 book *Cosmos* (New York: Random House), p. 43.

p. 223. The Viking spacecraft were not the first to reach the surface of Mars. Several Soviet probes crashed or attempted to land there in the late 1960s and early 1970s. None returned any meaningful data.

p. 225. For a summary of the early Martian climate debate, see J. K. Kasting, "The Early Mars Climate Question Heats Up," *Science* 278 (1997): 1245. The habitable-zone concept is discussed in G. Vogel, "Expanding the Habitable Zone," *Science* 286 (1999): 70–71.

p. 234. The first report of possible organic compounds in a Martian meteorite was I. P. Wright, M. M. Grady and C. T. Pillinger, "Organic Materials in a Martian Meteorite," *Nature* 340 (1989): 220–222. The paper on the Martian meteorite ALH84001 that generated such a controversy was D. S. McKay et al., "Search of Past Life on Mars: Possible Relic Biogenic Activity in Martian Meteorite ALH84001," *Science* 273 (1996): 924–930.

p. 237. Whether or not life has been exchanged among the planets in our Solar System in the past, we are going to start exporting life from Earth with a vengeance in the near future. We may already have done so inadvertently. Such contamination poses a real problem, raising questions about how easily we can distinguish the discovery of life on other worlds from contamination by our own.

In 1967, the United Nations passed the *Outer Space Treaty*, which dealt with the "principles governing the activities of states in the exploration and use of outer space, including the Moon and other celestial bodies." Article IX of this treaty was specifically concerned with contamination responsibilities of states during the new era of space exploration. The proposal of the Article was to establish an international policy requiring "that all space celestial bodies or adverse changes in the environment of the Earth from the introduction of extraterrestrial material."

"Planetary Protection" is of central concern as we begin to aggressively explore other bodies in the Solar System and retrieve extraterrestrial samples for study back here on Earth. Comprehensive policies have been developed by the Committee on Space Research (COSPAR) to guide NASA and other international space agencies. NASA has a full time planetary protection officer to assure mission compliance with these policies.

Spacecraft that land on other solar system bodies must meet certain "bioburden" requirements to insure we do not forward-contaminate these worlds with terrestrial organisms. In order to avoid back-contamination, any samples returned to Earth from solar system bodies that have even the remotest possibility that life might exist there now, or in the past, are required to be rigorously contained until they are proven not to present any type of hazard to terrestrial life.

When astronauts Pete Conrad, Richard Gordon, and Alan Bean, in *Apollo 12*, brought back pieces of the robot lander Surveyor 3 that had landed on the Moon three years earlier, bacteria were cultured from them. It is unclear whether the pieces of the robot had become contaminated as they were being returned, but if bacteria did manage to survive in the lunar environment for a substantial period, their survival stands as an incredible tribute to the toughness of life.

The potential for contamination of Mars is far greater than that for contamination of the Moon. The planetary protection requirements for spacecraft landing on solar system bodies on which there is even the remote chance of life reflect the difficulty of sterilization. Vehicles are supposed to carry fewer than 300,000 bacterial spores, which seems (and is) a lot.

Because the Viking landers carried life-detection experiments, extra precautions were taken to try to ensure that hitchhiking terrestrial organisms did not generate positive results that could have been due to contamination. The landers were subjected to a series of rigorous cleaning procedures. The Biology Instrument packages were sterilized by heating at 120°C for fifty-four hours before incorporation onto the landers. The landers themselves were encapsulated in a sealed bioshield that was later jettisoned in space, and were heated at 110°C for twenty to thirty hours.

It is estimated that all these precautions were very successful. Viking lander surfaces exposed to the Martian environment probably carried only about thirty bacterial spores. We can only hope that none of them were capable of multiplying on Mars.

Soviet spacecraft had crash-landed on the Martian surface before Viking. Did they comply with planetary protection standards? We now know that thorough sterilization was in fact carried out on those lander missions. Three identical spacecraft were built. Two were launched to Mars, where they crash-landed in 1971 and 1973 and failed to return any meaningful data. The third spacecraft, which remained on Earth, was checked for contamination in a very direct way indeed. It was smashed to pieces and completely ground to dust, which was then analyzed for viable organisms! The levels that were found were well within the limits imposed by the planetary protection guidelines. These Soviet tests were considerably more thorough than the ones that were carried out on the U.S. landers.

We can only hope that after all this, Mars is still pristine. But as we begin the new millennium, we are also beginning a new era of planetary protection concerns. Contamination of other planets and satellites by microorganisms from Earth may be avoidable, so long as we use sterilized robot probes. But are we prepared to protect Earth from the flood of samples that some of these probes will be bringing back?

So far, the only extraterrestrial samples that have been returned to Earth are from the Moon. But in the first two decades of the twenty-first century, a flotilla of spacecraft is being assembled to bring back samples from Mars, comets, asteroids, and possibly even Jupiter's moon Europa. Should we even try to obtain these samples until we are absolutely sure they are not hazardous?

The National Academy of Sciences and NASA have recommended that any testing of the returned Martian samples should be carried out in facilities similar to those used to test highly infectious agents such as the Ebola and Marburg viruses. Given all the recent public interest and scientific controversy about evidence for life in Mars meteorites, these recommendations are not too surprising.

These agencies have also recommended that many other samples be treated in the same way. These include samples that might be returned from Europa and its sister moon Ganymede, from Titan, from asteroids of the type that may have given rise to carbonaceous meteorites like Murchison, from comets, and even from the small Martian moons Phobos and Deimos. Life could be lurking on these bodies in a quiescent state, only to be awakened when brought to Earth.

Some have suggested using the International Space Station as a place to begin examining these samples. But there are many reasons why this will not work. It would require putting the probe that carries the samples into an orbit that matches that of the

NOTES AND REFERENCES
———

Space Station, which would be very difficult and extremely expensive. And even if scientists in the station do not break out in purple spots, this says nothing about what alien organisms might do if the scientists were to carry them inadvertently back to Earth. Unlike the Space Station, the planet Earth will provide these invaders with a full range of ecosystems to exploit.

Such strict containment recommendations may seem to be stretching risk avoidance to an extreme. If all these recommendations are to be followed, containment may be the single largest cost of a mission. Not surprisingly, there is a variety of opinions about the necessity of being so careful. Leslie Orgel, who has spent his career trying to reproduce life in the laboratory, has announced that he would gladly volunteer to eat portions of returned samples. This seems a safe bet, since organisms from other planets are unlikely to be adapted to his insides. But scientists cannot be cavalier about this—unless the public is satisfied that extraordinary safety measures are being taken, sample return missions will become highly controversial and support for bringing them back to Earth could quickly evaporate.

In short, bringing samples back from other planets may not be dangerous, but it will be perceived by the public as extremely dangerous. Exobiologists can only hope that the public will be satisfied with the efforts at containment currently being planned.

p. 239. A summary of the findings of the Galileo spacecraft observations of the Jupiter's four inner moons can be found in A. P. Showman and R. Malhotra, "The Galilean Satellites," *Science* 286 (1999): 77–84. For a discussion of oceans on Europa, see R. T. Pappalardo, J. W. Head, and R. Greeley, "The Hidden Ocean of Europa," *Scientific American*, October 1999: 54–63.

p. 244. For an account of what we know about Lake Vostok, see F. D. Carsey and J. C. Horvath, "The Lake That Time Forgot, *Scientific American,* October 1999: 62.

p. 246. For a discussion of extrasolar planets, see G. W. Marcy and R. P. Butler, "Giant Planets Orbiting Faraway Stars," *Scientific American Quarterly* 9 (1998): 10–15.

p. 250. The idea of rogue planets is discussed in D. Stevenson, "Life-Sustaining Planets in Interstellar Space?" *Nature* 400 (1999): 40.

p. 252. See C. Sagan et al., "A Search for Life on Earth from the Galileo Spacecraft," *Nature* 365 (1993): 715–721.

p. 254. For a discussion of the interferometry technique, see R. Angel and N. J. Woolf, "Searching for Life in Other Solar Systems," *Scientific American Quarterly* 9 (1998): 22–25.

Chapter 11: Epilogue

p. 257. For a discussion of the time required for the origin of life, see A. Lazcano and S. L. Miller, "How Long Did It Take for Life to Begin and Evolve to Cyanobacteria?" *Journal of Molecular Evolution* 39 (1994): 546–554.

INDEX

INDEX

INDEX

INDEX

INDEX

INDEX

INDEX

INDEX

INDEX